The Ultimate Guide to ChatGPT:

A beginner's handbook to understanding prompt engineering, the future of Artificial Intelligence and how to use it effectively.

Percival C. Verena

Disclaimer: The information and findings presented in this book are based on trial and error and are not necessarily supported by proven scientific fact. The purpose of this book is to provide insights and suggestions on how to use ChatGPT effectively, based on our own experiences and observations. We do not claim that our suggestions are the only way to use ChatGPT, nor do we guarantee that they will work for everyone. The results obtained may vary depending on the individual, their goals, and other factors. We strongly encourage readers to use ChatGPT for themselves to determine what works best for their goals and interests.

Table of Contents

Introduction

———————————◆‹‹◆››◆———————————

Artificial Intelligence (AI) has progressed by leaps and bounds since its creation. One of the most prominent applications of AI today is in the field of natural language processing (NLP). In addition, the development of language models has revolutionized how we interact with machines. As a result, the concept of "conversational AI" has become a buzzword in the tech industry. But what exactly is conversational AI, and how does it work?

Enter ChatGPT. ChatGPT is a conversational AI model developed by OpenAI that can generate human-like responses to a wide range of topics and questions. With its ability to understand and respond to natural language inputs, ChatGPT has the potential to transform the way we interact with machines and revolutionize various industries, from customer service to content creation.

However, while ChatGPT is a potent tool, it is crucial to understand its limitations and how to use it effectively to get the most out of it. This is where the concept of "prompt engineering" comes into play. Prompt engineering refers to the process of designing and shaping the conversation with ChatGPT through the use of well-crafted prompts.

This book aims to provide a comprehensive guide on how to effectively use ChatGPT and the importance of prompt engineering, as well as give some insight into how we can use ChatGPT in various industries for extreme efficiency and the betterment of everyone. We will cover everything from understanding the basics of ChatGPT and its capabilities to preparing for a conversation with ChatGPT, the importance of good prompts, and using ChatGPT for practical applications. Throughout the book, we will also explore the limitations of ChatGPT and how to overcome them, as well as the best practices for prompt engineering and conversation with ChatGPT.

This book is intended to introduce beginners to ChatGPT and show how powerful AI can be. Whether you are a tech enthusiast, a business professional, or a student, this book will provide valuable insights and practical tips to help you start using ChatGPT. We aim to make this book informative, entertaining, and easy to understand while providing specific examples to help illustrate key concepts.

So, whether you are looking to improve your AI skills or are curious about the potential of conversational AI, this book is for you. So, get ready to learn about the exciting world of ChatGPT!

CHAPTER 1

Introducing Chatgpt: An Advanced Conversational AI System

———————◆◀◀◆▶▶◆———————

The development of artificial intelligence has been a rapidly evolving field, and natural language processing is at the forefront of this growth. Chatbots and conversational AI systems have become increasingly popular and widespread, and ChatGPT is one of the latest and most advanced offerings in this space. In this chapter, we will take a closer look at ChatGPT, its history, and its capabilities to understand better what it is and can do.

GPT Language Model and its History

The GPT language model has become one of the most potent tools for natural language processing, capable of generating highly readable and human-like text. The history of GPT begins with the transformer architecture, which was first introduced in 2017 by Vaswani et al. Before the transformer, the most widely used models for NLP were based on recurrent neural networks (RNNs) and convolutional neural networks (CNNs). However, these models suffered from certain limitations, such as the difficulty of retaining long-term dependencies in the text.

The transformer architecture was designed to overcome these limitations by relying on a novel self-attention mechanism that allows the model to capture long-range dependencies more efficiently. This mechanism allows the model to focus on the essential parts of the input text and ignore the less relevant parts. The transformer architecture quickly gained popularity in the NLP community and became the basis for many state-of-the-art language models.

The GPT language model is based on the transformer architecture but with some additional modifications to make it more suitable for generating text. Specifically, the GPT model is trained in an unsupervised manner on massive amounts of text data to learn the patterns and relationships between words, sentences, and paragraphs. This training allows the model to generate new text that is highly coherent and consistent with the input it is given.

The first version of GPT was released in 2018, and it quickly gained attention in the research community for its impressive performance on various natural language processing tasks. Since then, several newer versions of GPT have been released, with each one improving on the performance of the previous one.

The GPT language model has been used for various applications, including language translation, summarization, chatbots, and more. It has also been used in research to generate realistic text for multiple applications, such as story writing and poetry. The ability of the GPT model to generate human-like text has sparked both excitement and concern in public, with some fearing that it may be used to generate fake news or other malicious content.

However, the GPT language model developers and the broader AI community have been working to ensure that the technology is used ethically and responsibly. For example, OpenAI has developed an API for GPT that allows developers to use the model for various applications while providing safeguards to prevent its misuse.

ChatGPT and its Capabilities

ChatGPT is a conversational AI system based on the GPT language model. It has been trained to generate human-like responses to various questions and prompts. This makes it an ideal tool for different use cases, including customer service, language translation, content generation, etc.

ChatGPT can understand and respond to natural language inputs, making it much easier to interact with than traditional chatbots requiring users to follow specific scripts or input patterns. Additionally, ChatGPT can generate human-like responses, allowing for more natural and engaging conversations.

Limitations and Challenges of Using ChatGPT

While ChatGPT is a powerful tool, it has limitations. One of the most significant limitations of the model is that it is only as good as the data it was trained on. If the training data contains biases or inaccuracies, the model will likely reflect them in its responses. Additionally, the model is imperfect and can sometimes generate nonsensical or inappropriate responses.

Another challenge of using ChatGPT is that it requires significant computational resources and may not be feasible for some users. Furthermore, the model can sometimes struggle with complex tasks or topics and may require human oversight to ensure the quality of its responses.

In addition to the limitations mentioned, ethical considerations must be taken into account when using ChatGPT. One such consideration is the potential for the model to be used for malicious purposes, such as generating fake news or propaganda. Another concern is ChatGPT's possibility to impersonate individuals or spread disinformation. To address these challenges, it is essential to ensure that ChatGPT is used responsibly and ethically. This includes regularly monitoring and updating the training data to prevent bias and inaccuracies, implementing safeguards to prevent malicious use, and providing human oversight to ensure the quality of responses. Despite these limitations and challenges, ChatGPT and other forms of AI have the potential to benefit society greatly. By recognizing and addressing these challenges, we can work towards creating a future in which AI is used responsibly and for the greater good.

ChatGPT vs Traditional Chatbots

ChatGPT is different from traditional chatbots in several key ways. Traditional chatbots often use rule-based systems or decision trees to generate responses. These systems are limited by the number of rules or paths they can follow and struggle with more complex conversations. In contrast, ChatGPT is based on a machine learning model trained on extensive text data. This allows it to generate more natural and flexible responses, making it well-suited for a broader range of use cases.

Traditional chatbots often require users to follow specific scripts or input patterns, limiting the range of possible conversations. ChatGPT, on the other hand, can interpret a wider variety of inputs and generate appropriate responses based on the context of the conversation. This makes it more user-friendly and can improve the overall user experience.

Another advantage of ChatGPT over traditional chatbots is its ability to learn and adapt over time. Traditional chatbots typically require manual updates and modifications to their rule-based systems, which can be time-consuming and labour-intensive. In contrast, ChatGPT can be continuously trained on new data, allowing it to improve its responses and adapt to user behavior or preferences changes. This makes it a more agile and effective solution for businesses and organizations seeking high-quality customer service or support.

Overall, while traditional chatbots have their place in certain use cases, ChatGPT significantly improves conversational AI technology. Its flexibility, natural language processing capabilities, and ability to learn and adapt make it a powerful tool for many applications, from customer service and support to content generation and language translation.

Natural language processing and conversational AI are rapidly advancing fields, with ChatGPT being one of the latest and most advanced offerings. The GPT language model, upon which ChatGPT is based, has become a powerful tool for natural language processing and has been used for various applications. ChatGPT is an ideal tool for different use cases, including customer service, language translation, content generation, and more.

CHAPTER 2

A Comprehensive Guide to Preparing for a Conversation with ChatGPT

———————◆‹‹◆››◆———————

Chatbots have been a part of our digital landscape for several decades now. However, with the advent of OpenAI's GPT language model, the capabilities of these chatbots have significantly evolved. The GPT language model is an artificial intelligence-powered language model capable of understanding natural language and generating human-like responses. As a result, ChatGPT, the chatbot powered by this language model, has become a valuable tool for a wide range of applications, from customer service to content creation.

However, the success of a conversation with ChatGPT depends on how well it is prepared. Preparing for a conversation with ChatGPT is known as prompt engineering. This practice involves designing and crafting prompts to guide the conversation towards a desired outcome. In this chapter, we will delve deeper into the importance of prompt engineering and provide a comprehensive guide to preparing for a discussion with ChatGPT.

The Importance of Understanding the Context of the Conversation

Understanding the context of the conversation is crucial because it allows you to tailor the conversation to the specific needs of your audience and the purpose of the discussion. For example, in the case of customer service, the conversation should focus on addressing the customer's needs and providing clear and concise responses that can help resolve their issue. Therefore, the prompt should be designed to elicit the necessary information from the customer, such as their problem or question, and guide the chatbot to provide relevant and helpful solutions.

Similarly, when the conversation is between a chatbot and a language model designed to generate creative content, the prompt should be designed to encourage the chatbot to generate unique and engaging content. This requires an understanding of the type of content that is expected and the target audience's preferences. The prompt should be designed to elicit imaginative, creative responses and tailored to the audience's specific needs.

Furthermore, understanding the environment in which the conversation will take place is essential because it can influence the tone and style of the discussion. For example, a conversation in a formal setting may require a more formal tone and structure, while a conversation in a casual setting may be more relaxed and informal. Therefore, it is essential to consider the context of the conversation when designing the prompt to ensure that it is appropriate for the setting and audience.

The Role of Prompt Engineering in Shaping the Conversation

Prompt engineering is a critical aspect of ensuring that the conversation with ChatGPT is effective and achieves the desired outcomes. A well-crafted prompt should be clear, concise, and relevant to the context of the conversation, as this will encourage ChatGPT to generate responses that align with the goals and objectives of the conversation. When designing prompts, it is essential to consider the conversation's specific context, including its purpose and intended audience.

For example, if the conversation is aimed at providing customer service, the prompts should be designed to encourage ChatGPT to provide clear and concise answers to customer questions. This will help ensure that the customer's needs are met and that the conversation remains focused on providing the necessary assistance. On the other hand, if the discussion is intended to generate creative content, the prompts should be designed to encourage ChatGPT to generate unique and engaging content. This will help ensure that the conversation remains creative and inspiring.

Overall, prompt engineering is crucial to shaping the conversation and guiding it towards a desired outcome. By crafting clear and relevant prompts that align with the context and goals of the conversation, you can help ensure that ChatGPT generates meaningful, relevant, and effective responses.

Key Factors to Consider When Preparing for a Conversation with ChatGPT

When preparing for a conversation with ChatGPT, several key factors must be considered. These include the length of the conversation, the type of conversation, and the complexity of the conversation.

The length of the conversation will determine the number of prompts required and the level of detail needed in each prompt. For example, a short discussion may only require a single prompt, while a longer conversation may require several prompts to guide the conversation.

The type of conversation will determine the tone and language of the prompts. For instance, a customer service conversation may require a friendly and helpful tone, while a content generation conversation may require a more creative and imaginative tone.

The complexity of the conversation will determine the level of detail required in each prompt. For instance, a customer service conversation may require clear and concise answers, while a content generation conversation may require more imaginative and creative responses.

Another key factor to consider when preparing for a conversation with ChatGPT is the structure of the conversation. A well-structured discussion can help ensure that the conversation remains focused on the desired outcomes and that ChatGPT can generate meaningful and relevant responses. The structure of the conversation can include the use of prompts, questions, and responses, as well as the overall flow.

In addition, it is essential to understand the context of the conversation, including the purpose of the discussion, the target audience, and the environment in which the conversation will take place. This information will help you design prompts tailored to the conversation's specific needs and the audience.

Finally, it is essential to be prepared to adapt the conversation as needed. While prompt engineering can help shape the conversation and guide it towards the desired outcomes, there may be times when the conversation takes unexpected turns. It is essential to be prepared to adapt and adjust the conversation as needed to remain focused on the desired outcomes. Considering these key factors, you can help ensure that your discussion with ChatGPT is productive, engaging, and effective.

Defining Clear Goals and Objectives for the Conversation

Before beginning a conversation with ChatGPT, defining clear goals and objectives is essential. This step is crucial in maximising the language model's capabilities.

Additionally, when preparing for a conversation with ChatGPT, it is essential to consider the language and terminology used. ChatGPT has been trained in a wide range of language and terminology, but it is necessary to understand its strengths and limitations to communicate with the model effectively. This includes being mindful of the language used in the prompt and ensuring that it is clear, concise, and relevant to the conversation context.

It is also essential to consider the tone of the conversation. The tone should be appropriate for the context and audience and consistent throughout the discussion. For example, a friendly and helpful tone is recommended if the conversation is for customer service purposes. On the other hand, if the conversation is for content creation, a more creative and imaginative tone may be more appropriate.

One key aspect of structuring a conversation with ChatGPT is using prompts, questions, and responses. By providing clear prompts and asking specific questions, you can help direct the conversation towards your desired outcomes and ensure that ChatGPT understands the direction of the conversation. Additionally, by offering thoughtful and detailed responses, you can help provide ChatGPT with the information it needs to generate relevant and valuable insights.

Another essential element of structuring a conversation with ChatGPT is the overall flow of the discussion. It's essential to consider the logical sequence of topics, ensuring that each point builds on the previous one and leads towards the desired conclusion. Additionally, it's essential to be aware of the pace of the conversation, allowing for enough time for ChatGPT to generate responses while keeping the conversation moving forward at an appropriate pace.

Be prepared to adapt the conversation as needed. While the immediate engineering process can help shape the conversation and guide it towards the desired outcomes, there may be times when the conversation takes unexpected turns. It is essential to be prepared to adapt and adjust the conversation as needed to remain focused on the desired outcomes. Adapting the conversation means being flexible and responsive to the conversation's direction. Even with careful planning and prompt engineering, there is always the possibility that the conversation may not go exactly as planned.

It also may involve changing the wording or tone of the prompts or questions being use or introducing new prompts or questions to redirect the conversation back towards the desired topic. It may also involve changing the overall structure or flow of the discussion in response to unexpected developments. It also requires a good understanding of the goals and outcomes of the conversation. By being prepared to adapt to the conversation, you can help ensure that you get the most out of your interactions with ChatGPT, even in situations where the conversation takes unexpected turns.

To wrap up, preparing for a conversation with ChatGPT is a critical step in maximizing the effectiveness of the language model and achieving the desired outcomes from the discussion. Understanding the context of the conversation, the role of prompt engineering, and the key factors to consider when preparing for a discussion with ChatGPT can help ensure that the conversation is effective and productive. By carefully considering the audience, the language and terminology used, the tone, the structure, and the ability to adapt as needed, you can ensure that the conversation with ChatGPT is successful and meets your goals and objectives.

CHAPTER 3

Crafting Chat-Worthy Prompts for ChatGPT: A Guide to Effective Conversations

——————◆◀◀◆▶▶◆——————

Welcome to the third chapter of our comprehensive guide to using ChatGPT. In this chapter, we will delve into the art of crafting the perfect prompts for ChatGPT, which is what truly sets it apart from other AI language models. As you've learned from the previous chapters, a good prompt can enhance your conversation and make ChatGPT perform like a seasoned comedian, while a poorly crafted one can lead to an awkward silence or a derailment of the conversation.

Clarity and Conciseness are Key

The first and most crucial rule in crafting effective prompts for ChatGPT is to keep them clear and concise. The more information you provide in the prompt, the easier it is for ChatGPT to understand what you're looking for and respond accordingly. Your prompt should act as a roadmap for the conversation, and the clearer the roadmap, the easier it is for ChatGPT to follow it and deliver a satisfying answer. For instance, instead of asking a general question like "What do you know about dinosaurs?" try a more specific one like "What are the characteristics of a Velociraptor?"

Another benefit of using clear and concise prompts is that they can help to prevent misunderstandings or irrelevant responses from ChatGPT. If the prompt is too vague or broad, ChatGPT may generate an answer that is not relevant to the intended conversation or may even provide incorrect information. In contrast, a specific and focused prompt helps to guide ChatGPT towards the desired outcome and increases the likelihood of generating a useful and accurate response.

Additionally, using clear and concise prompts can also help improve the conversation's overall efficiency. If the prompt is too long or complex, it may be difficult for ChatGPT to understand the intended meaning and respond promptly. On the other hand, a short and focused prompt can be easily understood by ChatGPT, allowing for a faster and more efficient conversation.

Tone is Everything

The second important component of a good prompt is the tone. Just like in human conversations, the tone of your prompt sets the tone for the entire conversation. If you want a lighthearted and playful conversation, use a playful and informal tone in your prompt. A more formal tone is appropriate if you're looking for more serious information. For example, if you want to know about the latest developments in space exploration, a playful prompt like "So, what's new in the world of space cowboys?" is likely to result in a lighthearted response, while a more formal prompt like "Please provide an update on the latest advancements in space exploration" will result in a more serious and informative response.

In addition to these examples, there are many other factors to consider when selecting the appropriate tone for your prompts. For instance, the type of audience you are targeting will play a significant role in determining the tone. A more casual tone may be appropriate when engaging with a younger audience, while a more professional tone may be required when engaging with a more mature audience or a corporate client.

Moreover, the context of the conversation is another important factor to consider when determining the tone. A prompt designed for a customer service conversation may require a friendly and approachable tone, while a prompt designed for a legal consultation may require a more formal and authoritative tone.

It's also important to be aware of potential biases or stereotypes that may be conveyed through the tone of your prompts. For instance, using a condescending or patronizing tone when addressing a female audience may be perceived as sexist or discriminatory.

Ultimately, the tone of your prompt should align with the goals and objectives of the conversation, as well as the preferences and expectations of your target audience. By carefully selecting the appropriate tone, you can create a more engaging and effective conversation with ChatGPT.

Language Matters

The language you use in your prompt also significantly impacts the response you receive from ChatGPT. ChatGPT may have trouble understanding what you're asking if you use overly technical or complex language. On the other hand, if you use slang or overly casual language, you may not get the level of detail you're looking for. It's best to aim for a balance between being clear and concise and appropriate for the tone you're trying to set. For example, if you're asking about the weather, a prompt like "Yo, what's the deal with the weather today?" is likely to result in a more casual response, while a prompt like "What is the current weather forecast?" will result in a more formal response.

Using appropriate language in your prompt is essential to ensure ChatGPT understands your message and responds appropriately. The language you use should match the context and tone of the conversation. For example, using colloquial language may be appropriate if you are having a casual conversation. However, if you're discussing a technical subject, using formal language may be necessary to ensure that ChatGPT can accurately understand and respond to your prompt.

Also, it's important to be mindful of cultural and linguistic differences. ChatGPT can be trained in a vast range of languages and dialects, but it's essential to make sure that the language used in the prompt is appropriate for the target audience. For instance, a prompt written in English may not translate well into another language, and ChatGPT may not fully understand certain cultural nuances.

Therefore, when crafting your prompt, it's important to consider the target audience and tailor the language used to ensure that ChatGPT can respond satisfactorily.

Guiding the Conversation with Prompts

Now that you understand the importance of clarity, tone, and language in crafting effective prompts, it's time to put that knowledge into practice. The ultimate goal of a good prompt is to guide the conversation in the desired direction. One of the most effective ways to do this is to ask open-ended questions that allow ChatGPT to take the conversation in various directions. For example, if you're curious about a certain topic, you might ask a question like "What do you think about [topic]?" This gives ChatGPT the flexibility to explore the topic in depth or provide a humorous take on it.

Additionally, you can also guide the conversation by using prompts that challenge ChatGPT to provide a specific perspective. For example, if you're interested in learning about the history of a particular event, you might ask, "What would have happened if [event] never took place?" This prompt challenges ChatGPT to consider alternate realities and provide a unique perspective.

Another effective way to guide the conversation is to use follow-up questions. By asking questions related to the response given by ChatGPT, you can keep the conversation flowing and steer it in a specific direction. For example, if you ask ChatGPT about the latest developments in space exploration and receive a response about a recent mission to

Mars, you could follow up with a question like "What was the mission's main goal to Mars?"

It's also important to remember that ChatGPT can understand context and build upon previous responses. By using prompts that are related to previous topics, you can create a conversation that flows naturally and seamlessly. For example, if you start a conversation about a specific movie and then ask a follow-up question about the director, ChatGPT will be able to understand the context and provide a response about the director of the movie.

As you can see, crafting the perfect prompt for ChatGPT is a combination of art and science. By paying attention to your prompts' clarity, tone, and language, you can create a conversational roadmap that will help guide ChatGPT in the desired direction. And by asking open-ended questions and following up with related prompts, you can keep the conversation flowing and build upon previous topics.

CHAPTER 4

Understanding the Limitations and Challenges of ChatGPT and the Importance of Human Intervention

————————◆‹‹◆››◆————————

The development of advanced language models such as ChatGPT has revolutionized the field of natural language processing, enabling computers to generate human-like responses to a wide range of prompts. However, despite its impressive capabilities, it is important to understand the limitations and challenges associated with using ChatGPT, as well as the importance of human oversight and intervention when using the model.

An Overview of ChatGPT Limitations

ChatGPT is trained on a large corpus of text data, but like any AI model, it is limited by the information it has been exposed to. ChatGPT may not have access to the most current information, and its responses may be limited by the data it has been trained on. For example, if ChatGPT was trained on data from 2018, it may not have information on events or developments that have occurred since that time.

Additionally, ChatGPT may struggle with tasks that require a deep understanding of context or abstract reasoning. For example, it may have difficulty answering questions that require common sense knowledge, such as "What is the capital of the United States?" or "What is the most common type of tree in a forest?" These limitations highlight the need for caution when using ChatGPT for certain tasks, particularly those that require a deep understanding of language and context.

The Challenges of Using ChatGPT for Complex Tasks

Despite its limitations, ChatGPT can be a powerful tool for various applications, but it may struggle with more complex tasks that require a deeper understanding of language and context. For example, ChatGPT may struggle with tasks that require a high level of detail or precision, such as writing a detailed scientific paper or composing a legal document. This can be due to the complexity of the language used in these types of tasks, as well as the need for a deep understanding of the subject matter.

Moreover, ChatGPT may have difficulty understanding and generating responses to questions that require cultural or societal context. For example, a prompt like "What is the best way to approach a coworker about a sensitive issue?" may result in an insensitive or inappropriate response due to the lack of understanding of cultural norms and societal expectations. This highlights the importance of considering the cultural and societal context when using ChatGPT for certain tasks.

The Importance of Human Oversight and Intervention

Given the limitations and challenges of using ChatGPT, it is important to have human oversight and intervention when using the model. This can help ensure that the responses generated by ChatGPT are accurate, appropriate, and aligned with the intended use case.

One way to achieve this is by using a human-in-the-loop approach, where a human is involved in the decision-making process for each response generated by ChatGPT. This can help address any potential biases or inaccuracies in the responses generated by the model and ensure that the final output is of high quality.

For example, in a customer service chatbot powered by ChatGPT, a human operator may review the responses generated by the model and make any necessary adjustments before they are sent to the customer. This can help ensure that the responses are customer-centric and aligned with the overall customer experience and can help address any cultural or societal sensitivities that may arise in the conversation.

The Role of Human-in-the-Loop Approaches in Ensuring Quality

Human-in-the-loop approaches play a crucial role in ensuring the quality and accuracy of responses generated by ChatGPT. These approaches help address the limitations and challenges of the model and ensure that the final output human-in-the-loop approaches play in ensuring the quality of responses generated by the model. While ChatGPT can generate human-like responses, it is limited by the data it has been trained on and may struggle with complex tasks that require a deep understanding of context and abstract reasoning.

In order to address these limitations, it is essential to have human oversight and intervention in the decision-making process for each response generated by ChatGPT. This helps to mitigate any biases or inaccuracies in the responses, and ensures that the final output aligns with the intended use case.

Additionally, human-in-the-loop approaches can help to enhance the quality of responses generated by ChatGPT. For example, in a customer service chatbot powered by ChatGPT, a human operator can review the responses generated by the model and make any necessary adjustments to ensure that they are customer-centric and aligned with the overall customer experience.

As ChatGPT continues to be developed and deployed in various applications, it is essential to consider the limitations and challenges of the model and the role that human oversight and intervention play in ensuring the quality of its responses.

One way to further address the limitations and challenges of ChatGPT is through ongoing training and refinement of the model. This can involve exposing the model to additional data and fine-tuning its parameters to improve its performance on specific tasks. However, it is essential to recognize that even with ongoing training and refinement, ChatGPT may still struggle with certain tasks that require a deep understanding of context and abstract reasoning.

Another way to address the limitations and challenges of ChatGPT is through complementary AI models. For example, a customer service chatbot powered by ChatGPT could also be integrated with a sentiment analysis model to help ensure that the responses generated are appropriate and aligned with the overall customer experience.

While ChatGPT is a powerful tool for generating human-like responses, it is important to understand its limitations and challenges and the role that human oversight and intervention play in ensuring the quality of its responses. Ongoing training and refinement of the model, as well as the use of complementary AI models, can help to address its limitations and enhance its performance.

CHAPTER 5

Enhancing the Conversation with ChatGPT

————————◆ ◀◀◆▶▶ ◆————————

C hat GPT can generate high-quality responses to user prompts, but several techniques can be used to further enhance the conversation and improve the quality of its responses. Let's explore the various methods for enhancing the conversation with ChatGPT and provide examples and detail on how to implement said techniques to achieve the best results.

Context-Aware Prompts

The prompt used to initiate a conversation with ChatGPT plays a crucial role in shaping the direction of the conversation and determining the quality of the responses generated by the chatbot. A well-crafted prompt can provide ChatGPT with important context and help to generate more personalized and relevant responses.

Context-aware prompts include additional information about the user or the context of the conversation. For example, a prompt that includes the user's location, interests, or previous interactions with the chatbot can provide ChatGPT with additional context that can be used to generate more personalized and relevant responses.

For instance, if a user is speaking with a weather chatbot powered by ChatGPT, the prompt may include the user's location in order to provide them with accurate weather information. In this case, the context-aware prompt can help the chatbot generate a response tailored to the user's location, such as "The current temperature in Los Angeles is 72°F."

Another example of a context-aware prompt could consider the user's previous interactions with the chatbot. For instance, if the user had previously asked the weather chatbot about the forecast for the next day, a context-aware prompt could be "What is the forecast for tomorrow in Los Angeles?" This prompt considers the user's previous interaction and provides additional context, which can help the chatbot generate a more personalized and relevant response. Context-aware prompts can also be used to guide the conversation in a specific direction. For example, if the user is interested in a particular topic, a context-aware prompt could be crafted to steer the conversation towards that topic. This can be achieved by including relevant keywords or phrases in the prompt that signal the user's interest. For instance, if the user is interested in learning about new movies, a context-aware prompt could be "What are some of the latest movie releases?" This prompt provides the chatbot with meaningful context and signals the user's interest, which can help to generate a more engaging and relevant response.

Pre-Defined Templates

Pre-defined templates can be an effective way to improve the conversation flow and accuracy of ChatGPT responses. These templates can be tailored to specific use cases and provide a consistent chatbot interaction structure. Pre-defined templates can be particularly useful when the chatbot needs to gather specific information or guide the conversation towards a particular outcome.

One example of a pre-defined template is a booking template used by travel chatbots. This template includes steps for gathering information such as the traveler's destination, travel dates, and preferred airline. The chatbot can then use this information to search for available flights and provide the user with options that match their preferences.

Another example of a pre-defined template is a customer service template, as mentioned earlier. This template can include steps for identifying the customer's issue, gathering relevant information, and providing a resolution. The chatbot can then use this information to provide a suitable response and guide the conversation towards a resolution.

Pre-defined templates can also be combined with other techniques, such as natural language processing and machine learning, to improve the accuracy and relevance of responses. For instance, a travel chatbot may use natural language processing to understand the user's query, and then use a pre-defined template to guide the conversation towards booking a flight. The chatbot can then use machine learning to improve its understanding of the user's preferences and provide more personalized recommendations in future interactions.

Different Forms of Input

In addition to text-based prompts, ChatGPT can also be used to generate responses to different input forms, such as images or videos. The use of various forms of input can provide ChatGPT with additional context and help to shape the direction of the conversation.

For example, an image recognition chatbot powered by ChatGPT may use an image as input to generate a response that describes the image or provides information about the objects in the image. The chatbot may use image recognition technology to identify objects in the image and generate a response like "I see a beautiful sunset over the ocean in the image."

Similarly, a video recognition chatbot powered by ChatGPT may use a video as input to generate a response that describes the content of the video. For instance, the chatbot may use video recognition technology to identify a sports event in a video and generate a response like "I see a basketball game in progress in the video."

The ability of ChatGPT to generate responses to different forms of input allows it to be used in various domains and enhances its capability to understand and respond to users' needs.

Reinforcement Learning

Reinforcement learning can also be combined with other techniques, such as context-aware prompts and pre-defined templates, to provide an even more powerful and comprehensive approach to shaping the conversation with ChatGPT. For example, a customer service chatbot powered by ChatGPT may use a combination of context-aware prompts and reinforcement learning techniques to provide a highly personalized and relevant experience for the customer. The chatbot may prompt the customer for their location and previous interactions, use this information to determine the customer's preferences, and then use reinforcement learning techniques to further shape the conversation based on the customer's feedback.

One important thing to remember when using reinforcement learning is to ensure that the feedback the user provides is accurate and relevant. For example, if the customer provides feedback that is not related to the conversation or is not representative of their true preferences, the reinforcement learning algorithms may not be able to shape the conversation effectively. To address this, it may be necessary to implement additional techniques, such as sentiment analysis or natural language processing, to help ensure that the feedback provided by the user is accurate and relevant.

Another important aspect to consider when using reinforcement learning is the training process. It is important to ensure that the chatbot is trained on a large and diverse set of data, which can help to improve the accuracy and quality of the chatbot's responses. The training process should also be regularly monitored and updated as new data becomes available or as the user's preferences and behaviors change over time.

There are many techniques that can be used to enhance the conversation with ChatGPT and improve the quality of its responses. Using context-aware prompts, pre-defined templates, different forms of input, and reinforcement learning can help shape the conversation and provide ChatGPT with the necessary context to generate high-quality responses. By implementing these techniques and carefully considering the feedback provided by the user, it is possible to create highly personalized and context-aware chatbots that can provide an exceptional user experience.

Most of these techniques are available with the current ChatGPT version, however, future versions will allow for much more specific and complex techniques to be implemented.

CHAPTER 6

Using ChatGPT for Practical Applications

————————◆◄◆►◆————————

As a large language model developed by OpenAI, ChatGPT has a wide range of potential applications, from customer service to language translation and content generation. In this chapter, we will explore the different use cases for ChatGPT and its benefits and challenges for each of these applications. We will also examine the best practices for using ChatGPT in each use case and discuss the future potential of the technology and its role in shaping the future of artificial intelligence.

Content Generation

Another potential use case for ChatGPT is a content generation, such as generating articles, blog posts, and other types of written content. ChatGPT can be used to generate high-quality written content in real-time, providing businesses and individuals with a cost-effective and efficient solution for content creation.

For example, a content generation chatbot powered by ChatGPT can be used to generate articles, blog posts, and other types of written content. The chatbot can be trained to generate content on various topics, from technology and business to sports and entertainment. Using pre-defined templates and context-aware prompts can help shape the direction of the content and ensure that it is relevant and appropriate.

One of the challenges of using ChatGPT for content generation is ensuring that the chatbot provides high-quality content that is engaging and relevant to the target audience.

To overcome this challenge, training ChatGPT on a large and diverse corpus of high-quality text data is important and fine-tuning the model for specific use cases is important. Additionally, using context-aware prompts and templates can help shape the content's direction and ensure that it is tailored to the target audience. It's also important to carefully review and edit the content generated by the chatbot to ensure that it is error-free and meets the desired standards of quality.

Furthermore, there are potential ethical considerations when using ChatGPT for content generation, particularly when it comes to issues such as plagiarism and copyright infringement. It's important to ensure that any chatbot-generated content does not violate intellectual property rights or ethical standards.

Overall, content generation is a promising use case for ChatGPT, offering businesses and individuals a cost-effective and efficient solution for creating high-quality written content. However, it's important to consider the potential ethical implications carefully and to ensure that the chatbot's content meets the desired standards of quality and relevance.

ChatGPT for Customer Service

We have covered this in earlier chapters a bit, but ChatGPT has the potential to revolutionize the customer service industry by providing businesses with a cost-effective and highly efficient solution for handling customer inquiries. Using ChatGPT in customer service can provide customers with quick and accurate answers to their questions, which can help improve the overall customer experience.

One of the key benefits of using ChatGPT for customer service is its ability to handle large volumes of customer inquiries. ChatGPT can be configured to handle multiple customer inquiries simultaneously, which means that businesses can handle more inquiries with the same resources. Additionally, ChatGPT's natural language processing capabilities allow it to understand and respond to customer inquiries in a human-like manner, which can help to improve the overall customer experience.

However, some challenges are associated with using ChatGPT for customer service. One of the biggest challenges is ensuring that ChatGPT provides accurate and relevant responses to customer inquiries. Businesses must ensure that their ChatGPT model is well-trained and up-to-date with the latest information in order to provide customers with accurate and relevant responses. Additionally, businesses must ensure that their ChatGPT model is configured to handle different types of customer inquiries, such as complex technical issues or sensitive customer information.

Best Practices for Using ChatGPT in Customer Service

In order to get the most out of using ChatGPT for customer service, businesses should follow a few best practices:

1. Invest in quality training data: In order to provide accurate and relevant responses, it is important to invest in quality training data. This means providing ChatGPT with a large and diverse set of customer inquiries, as well as information about the business and its products and services.

2. Regularly update the ChatGPT model: The information and data used to train the ChatGPT model will become outdated over time. As a result, it is important to regularly update the ChatGPT model with the latest information to ensure that it provides accurate and relevant responses.

3. Monitor customer feedback: Monitoring customer feedback can help businesses to identify areas where ChatGPT is providing incorrect or irrelevant responses, which can then be corrected in the model.

4. Integrate with other customer service tools: ChatGPT can be integrated with other customer services tools, such as live chat and voice assistants, to provide customers with a seamless and comprehensive customer service experience.

ChatGPT for Language Translation

ChatGPT also has the potential to revolutionize the language translation industry by providing businesses with a highly accurate and cost-effective solution for translating text and speech. Using ChatGPT in language translation can help businesses expand into new markets and reach a larger customer base.

One of the key benefits of using ChatGPT for language translation is its ability to provide highly accurate translations. ChatGPT's natural language processing capabilities allow it to understand and translate text and speech in a manner that is more accurate and natural than traditional language translation tools.

However, some challenges are associated with using ChatGPT for language translation. One of the biggest challenges is ensuring that ChatGPT can handle different languages and dialects. Businesses must ensure that their ChatGPT model is well-trained and up-to-date with the latest information for the specific languages and dialects they translate.

Finally, it's essential to consider the future potential of ChatGPT and its role in shaping the future of AI. As ChatGPT continues to evolve and improve, it has the potential to be used for a wide range of new and innovative applications. For example, ChatGPT could generate highly personalized virtual assistants, provide real-time translation services, or even create new content forms, such as news articles or short stories.

The future of ChatGPT is also closely tied to the development of other AI technologies, such as natural language processing and computer vision. As these technologies continue to improve, ChatGPT can better understand and respond to a wide range of input, from text and speech to images and videos.

Moreover, the use of reinforcement learning and other advanced machine learning techniques will play a critical role in shaping the future of ChatGPT. As ChatGPT continues to learn from interactions with users, it will be able to improve its responses and generate more personalized and relevant responses over time.

In conclusion, ChatGPT is a powerful tool with many practical applications, from customer service and language translation to content generation. By understanding the benefits and challenges of using ChatGPT for each of these use cases, as well as best practices for implementation, organizations can unlock the full potential of ChatGPT and use it to deliver a more personalized and engaging user experience. With its future potential and role in shaping the future of AI, ChatGPT has the potential to revolutionize how we interact with technology and the world around us.

CHAPTER 7

Best Practices for Prompt Engineering and Conversation with ChatGPT

————————◆ ◀◀ ◆ ▶▶ ◆————————

Chat GPT has the potential to revolutionize the way we communicate and interact with technology, but to realize its potential fully, it is essential to understand the key principles and best practices for effective, prompt engineering and conversation with ChatGPT. In this chapter, we will explore the key elements that go into designing effective prompts and shaping the conversation with ChatGPT.

Key Principles of Prompt Engineering

Prompt engineering is a critical component of the conversation with ChatGPT. The prompt's quality can greatly impact the response generated by ChatGPT, and it is essential to understand the key principles of effective, prompt engineering. Some of the key principles of prompt engineering include:

- Providing clear and concise information about the context of the conversation.

- Providing additional information about the user or the topic of the conversation.

- Providing a structure or template for the conversation can help ChatGPT understand the context and generate more relevant and personalized responses.

By following these principles, you can design effective prompts that provide ChatGPT with the information it needs to generate high-quality responses.

Monitoring and Evaluating the Conversation

Monitoring and evaluating the conversation with ChatGPT is an essential part of the process of shaping the conversation. This can help you to identify areas where the conversation can be improved and to make changes to the prompt or the response generation process as needed.

Several different tools and techniques can be used to monitor and evaluate the conversation with ChatGPT. Some of these include:

- Logging and analyzing the conversation data can help you understand the patterns and trends in the conversation.

- Using metrics, such as response accuracy, response relevance, and customer satisfaction, to measure the quality of the conversation.

- Collecting and analyzing feedback from users, which can help you to identify areas where the conversation can be improved.

By monitoring and evaluating the conversation with ChatGPT, you can gain a deeper understanding of the strengths and weaknesses of the system and make changes to improve the quality of the discussion over time.

Using Data and Feedback to Improve the Conversation

Data and feedback are critical in shaping the conversation with ChatGPT. By analyzing the data generated by the discussion and the feedback received from users, you can gain a deeper understanding of the strengths and weaknesses of the system and make changes to improve the quality of the conversation over time.

Some of the ways that you can use data and feedback to improve the conversation with ChatGPT include:

- Using data to identify patterns and trends in the conversation can help you identify areas where the conversation can be improved.

- Using user feedback to identify areas where the conversation is not meeting their expectations and making changes to the prompt or response generation process as needed.

- Using data to understand the user's preferences and make changes to the prompt or response generation process to provide more personalized and relevant responses.

By using data and feedback to improve the conversation with ChatGPT, you can create a more engaging and personalized conversation that meets the needs and expectations of your users.

Human Judgment and Ethical Considerations

In addition to the technical aspects of using ChatGPT, there are also important ethical considerations to take into account. The use of ChatGPT can raise questions about the role of humans in shaping the conversation and the extent to which ChatGPT is capable of making decisions on its own.

To ensure that the use of ChatGPT aligns with ethical principles, it is essential to consider the potential consequences of the chatbot's responses and ensure that the chatbot is not used to spread false information or engage in unethical behavior. For example, a chatbot powered by ChatGPT used for customer service should be programmed to follow ethical customer service practices and avoid spreading false information to customers.

Another important ethical consideration is the potential for ChatGPT to perpetuate existing biases and reinforce stereotypes. For example, a chatbot powered by ChatGPT that is trained on biased data may generate responses that perpetuate existing biases and reinforce stereotypes. To minimize the potential for bias, it is important to carefully curate the training data used to train the chatbot and ensure that the chatbot is trained on diverse and inclusive data.

Human judgment is essential when evaluating the quality and appropriateness of ChatGPT's responses. While ChatGPT can generate responses quickly and efficiently, it may not always provide the most appropriate or accurate responses. Humans need to review the chatbot's responses and ensure that they are aligned with the intended purpose of the chatbot and that they are appropriate for the context in which they are being used.

Furthermore, as with any technology, there is the potential for ChatGPT to be misused or abused. It is essential for organizations and individuals to use ChatGPT responsibly and ethically and to ensure that the chatbot is not used to spread false information, engage in unethical behavior, or violate users' privacy. Overall, ChatGPT requires a balance between technical expertise, ethical considerations, and human judgment. By taking into account these factors, organizations and individuals can ensure that they are using ChatGPT responsibly and effectively.

Here are nine ChatGPT prompts that will accelerate your learning.

1) Prompt: "Help me draw parallels between (topic or skill) and unrelated domains, promoting a deeper understanding and the ability to apply knowledge in novel situations." – use analogical thinking for better comprehension.

2) Prompt: "Develop a lesson plan for me to teach (topic or skill) to a friend or colleague, reinforcing my understanding and identifying areas for improvement." – encourage teaching others as a learning strategy.

3) Prompt: "Guide me on how to formulate effective questions related to (topic or skill) that will promote deeper thinking and stimulate curiosity." – learn to ask better questions.

4) Prompt: "Create a learning plan that incorporates real-world scenarios or applications for (topic or skill) to enhance memory retention and understanding." – utilize the power of context for improved recall.

5) Prompt: "Break down (topic or skill) into smaller, more manageable chunks or units, and provide a structured plan for mastering each one." – apply the concept of "chunking" for easier learning.

6) Prompt: "Suggest ways to incorporate the learning of (topic or skill) into my daily routine, creating habits that support consistent progress and growth." – apply the power of habit formation to learning.

7) Prompt: "Provide me with a series of quizzes or tests on (topic or skill) at varying difficulty levels to enhance my retention and understanding." – leverage the testing effect for better retention.

8) Prompt: "Create a study plan incorporating the Pomodoro Technique to help me maintain focus and productivity while learning (topic or skill)." – Utilize the Pomodoro Technique for focused learning.

9) Prompt: "Introduce challenges or obstacles related to (topic or skill) that will force me to think more deeply and enhance my learning and problem-solving abilities." – Utilize the concept of "desirable difficulties."

Using ChatGPT for practical applications can bring many benefits. By monitoring and evaluating the conversation, using data and feedback to improve the discussion over time, and considering the role of human judgment and using hyper-specific prompts, ChatGPT can be used to generate high-quality responses and provide the user with limitless information and assistance.

CHAPTER 8

Advanced Techniques for Improving ChatGPT Performance

———————— ◆ ‹‹ ◆ ›› ◆ ————————

C hat GPT is a powerful language generation tool that can produce high-quality and context-aware responses. However, as with any AI system, there are many advanced techniques that can be used to improve the performance of ChatGPT further. In this chapter, we will explore some of the most effective methods for enhancing the performance of ChatGPT and increasing the quality of its responses.

Fine-Tuning Pre-Trained Models

Fine-tuning pre-trained models is an essential technique for enhancing ChatGPT's performance. Fine-tuning involves training the model on a specific task or dataset, allowing it to better understand the context and language of the task. This approach can be particularly useful when working within a particular domain, such as customer service or language translation, as it enables ChatGPT to become more specialized and efficient in that field.

Transfer learning is the method used to fine-tune pre-trained models, which involves leveraging the pre-trained weights of a model and updating them with task-specific data. This allows the model to adapt to the new data and improve its performance while still benefiting from the knowledge gained from the pre-trained model. Transfer learning helps reduce the time and resources required to train a model from scratch and enables ChatGPT to handle new tasks and domains better.

Fine-tuning can also help mitigate some of ChatGPT, such as its lack of knowledge on specific domains or its inability to understand the context and common sense knowledge. ChatGPT can generate more accurate and relevant responses for specific tasks and domains by fine-tuning the model on task-specific data and providing context-aware prompts. However, ensuring that the training data used for fine-tuning is diverse, inclusive, and free from biases is essential to avoid perpetuating existing biases and stereotypes.

Data Augmentation

Data augmentation is another powerful technique for improving the performance of ChatGPT. This process involves creating additional training data by artificially transforming existing data. For example, data augmentation could involve randomly changing the order of words in a sentence, adding noise to the data, or creating synthetic data through text generation techniques.

Data augmentation can help to reduce overfitting, which is when a model becomes too specialized to the training data and performs poorly on new data. ChatGPT can better generalize to new data and produce higher-quality responses by creating additional data through data augmentation.

Expanding on the previous point, data augmentation is a technique that can help to increase the size and diversity of the training data used to train ChatGPT. This is important because larger and more diverse training data can help ChatGPT better understand language nuances and improve its ability to generate high-quality responses.

There are several different methods of data augmentation that can be used with ChatGPT. One common method is adding noise to the data, such as typos or misspellings to the text. This can help to simulate the variability of natural language and improve the ability of ChatGPT to handle noisy or incomplete input.

Another method of data augmentation is to modify the input data randomly. This could involve randomly swapping words or phrases, adding or removing words, or changing the order of words in a sentence.

By making these changes, ChatGPT can be exposed to a wider range of input data and better generalize to new situations. In addition to these methods, data augmentation can also involve generating synthetic data using techniques such as text generation. This can be especially useful in cases where there is a limited amount of training data available, as it allows ChatGPT to generate additional data and improve its performance.

Multi-Task Learning

Multi-task learning is a powerful technique that can be used to improve the performance of ChatGPT. This approach involves training a single model on multiple tasks simultaneously rather than training separate models for each task. The goal is to allow the model to leverage the knowledge and experience gained from one task to improve performance on the others.

For example, a ChatGPT model trained on customer service data could also be trained on language translation data. By training the model on both tasks, it can learn to understand better the nuances of customer service interactions and the intricacies of language translation. The knowledge gained from one task can be used to improve the performance of the other, resulting in higher-quality responses for both tasks.

Multi-task learning can also help reduce the training data needed for each task. By sharing knowledge across tasks, the model can become more efficient at learning and require less data to achieve high performance.

However, there are some challenges associated with multi-task learning. For example, the tasks being learned must be related in some way so that knowledge can be transferred between them. Additionally, the model may become more complex and require more resources to train and deploy.

Despite these challenges, multi-task learning is a promising approach for improving the performance of ChatGPT and can lead to more efficient and effective chatbots.

Ensemble Methods

Ensemble methods are a popular technique for boosting the performance of machine learning models, including ChatGPT. Instead of relying on a single model to make predictions, ensemble methods use multiple models to generate predictions and then combine them to produce a more accurate result.

There are different ways to implement ensemble methods in ChatGPT, such as training multiple models on different subsets of data, using different architectures or hyperparameters, or fine-tuning models for specific tasks. The final prediction is then made by aggregating the predictions from each model using various techniques, such as weighted average, max voting, or stacking.

They also can help address individual models' limitations and improve the overall performance of ChatGPT. Combining the predictions of multiple models, ensemble methods can reduce the risk of overfitting, increase the model's ability to generalize to new data, and enhance the model's robustness to noisy or ambiguous input. Ensemble methods can also help to capture different aspects of the data and generate more diverse responses, which can be particularly useful in natural language processing applications like chatbots.

However, ensemble methods also require additional computational resources and can be more complex to implement and maintain than individual models. Careful selection of the models to include in the ensemble and the method used to aggregate the predictions is crucial to ensure optimal performance.

Continual Learning

Continual learning is a technique that involves training a model to continuously learn from new data without forgetting the knowledge gained from previous data. This can be especially useful for ChatGPT, as it allows the model to continuously improve its performance over time and better adapt to changing data distributions.

For example, a ChatGPT model could be trained on customer service data, and then continuously updated with new customer service data. The model would then be able to learn from the new data and improve its performance over time without forgetting the knowledge gained from the previous data.

In conclusion, these advanced techniques for improving the performance of ChatGPT can be highly effective in increasing the quality of its responses and better adapting to specific tasks or domains. Whether through fine-tuning pre-trained models, data augmentation, multi-task learning, ensemble methods, or continual learning, these techniques offer the opportunity to take ChatGPT to the next level and realize its full potential.

However, it is essential to remember that while these techniques can be highly effective, they also require a significant amount of technical expertise and resources to implement. Therefore, it is recommended to carefully consider the goals and resources available before embarking on any advanced techniques for improving ChatGPT performance. Additionally, it is essential to monitor and evaluate the performance of ChatGPT continually, and to make adjustments as needed to ensure that it is meeting the desired goals and standards.

CHAPTER 9

Understanding and Interpreting ChatGPT Output

————◆◄◆►◆————

hat GPT can generate highly convincing and context-aware responses, making it an attractive tool for many applications. However, it is essential to understand that the responses generated by ChatGPT are not always accurate, and it is crucial to be able to evaluate and interpret the output from ChatGPT.

In this chapter, we will explore the different types of output that can be generated by ChatGPT and discuss how to evaluate the accuracy and quality of these responses effectively. We will also discuss the importance of monitoring and evaluating the performance of ChatGPT and the role that human judgment plays in interpreting its output.

Different Types of ChatGPT Output

ChatGPT can generate a wide range of output, including natural language text, images, and audio. When working with ChatGPT, it is essential to understand the different types of output that can be generated, as this can impact the accuracy and quality of the responses.

Text-based Output: Text-based output is the most common type of output generated by ChatGPT and is used for a wide range of applications, including customer service, language translation, and content generation. This type of output is generated by processing natural language inputs and generating natural language responses.

Image-based Output: Image-based output is generated by processing image data and generating new images. This output type is used for image classification, image generation, and image captioning applications.

Audio-based Output: Audio-based output is generated by processing audio data and generating new audio. This output type is used for applications such as speech synthesis and music generation.

In addition to the three main types of output mentioned above, there are also variations and combinations of these output types that ChatGPT can generate. For example, the text-based output can be combined with image-based output to generate image captions or with audio-based output to generate speech-to-text transcripts.

Moreover, ChatGPT can generate output in different languages, depending on the language of the input and the language model used for training. ChatGPT can support multilingual conversations and expand its applications to global markets.

The type of output generated by ChatGPT is largely dependent on the task or application it is being used for. For instance, in customer service applications, text-based output is typically used to generate responses to customer queries, while image-based output may be used in applications such as facial recognition.

Understanding the different output types that ChatGPT can generate is essential for developers and businesses to determine the most suitable output format for their specific applications or task.

Evaluating the Accuracy and Quality of ChatGPT Output

When working with ChatGPT, it's essential to evaluate the accuracy and quality of the responses generated by the model. There are several metrics that can be used to assess the performance of the model.

Precision: Precision is a metric that measures the proportion of relevant responses generated by ChatGPT. It helps to evaluate the model's ability to generate relevant and accurate responses. A high precision score indicates that the model is generating highly relevant responses to the input prompt.

Recall: Recall is a metric that measures the proportion of relevant responses that ChatGPT correctly generates. It evaluates the model's ability to capture all of the relevant information in the input prompt. A high recall score indicates that the model effectively captures all of the important information in the prompt.

F1-Score: The F1-score is a metric that balances precision and recall. It's a good overall measure of the model's performance. A high F1-score indicates that the model is generating accurate and relevant responses.

Perplexity: Perplexity is a metric that evaluates the model's ability to predict the next word in a sequence. It assesses the model's ability to generate coherent and grammatically correct responses. A low perplexity score indicates that the model generates fluent and natural-sounding responses.

It's essential to keep in mind that no single metric can fully evaluate the performance of ChatGPT. Therefore, a combination of metrics is recommended to assess the accuracy and quality of the model's responses. Additionally, human experts' manual evaluation of the reactions generated by ChatGPT can provide valuable insights into the model's performance.

Monitoring and Evaluating the Performance of ChatGPT

Monitoring and evaluating the performance of ChatGPT is an ongoing process that helps to ensure that the model continues to generate high-quality and relevant responses. There are several techniques that can be used to monitor and evaluate the performance of ChatGPT:

1. Regular Performance Evaluations: Regular performance evaluations are an essential tool for identifying areas where the model is performing well and where there is room for improvement. These evaluations can be conducted by comparing the model's responses to human-generated

responses, evaluating the model's accuracy and precision, and analyzing the model's performance over time.

2. Monitoring Model Outputs: Monitoring model outputs is critical for detecting model behaviour changes. This can be done by analyzing the model output for consistency and accuracy, tracking the model's response time, and monitoring the model's resource utilization.

3. Feedback from End-Users: Feedback from end-users is an invaluable tool for evaluating the quality and relevance of the model's responses. End-users can provide feedback through surveys, ratings, and reviews, which can be used to identify areas where the model is performing well and where there is room for improvement.

4. A/B Testing: A/B testing is a technique used to compare the performance of different models or variations of the same model. This technique involves randomly assigning users to other models and measuring the performance of each model. This can help identify which model generates the most accurate and relevant responses.

5. Continuous Learning: Continuous learning is updating and improving the model over time. This can be done by incorporating new data into the model, refining the model's algorithms, and updating the model's training data. Continuous learning ensures that the model remains up-to-date and generates high-quality responses over time.

Regular performance evaluations, monitoring of model outputs, and feedback from end-users are all valuable techniques that ensure that the model continues to generate high-quality and relevant responses over time. By implementing these techniques, businesses and developers can ensure that their ChatGPT-powered chatbots and conversational agents remain effective and reliable tools for engaging with customers and users.

The Role of Human Judgment in Interpreting ChatGPT Output

While ChatGPT is a powerful tool for generating high-quality and context-aware responses, it is essential to acknowledge that the responses generated by the model are not always perfect and can sometimes contain inaccuracies or biases. As a result, human judgment must play a role in interpreting the output from ChatGPT to ensure that the responses generated are relevant, trustworthy, and accurate.

There are many factors that can impact the accuracy of ChatGPT's output, including the quality of the data used to train the model, the complexity of the input, and the limitations of the model's architecture. As a result, human judgment needs to be used to assess the quality and accuracy of ChatGPT's output and to make any necessary corrections or modifications.

In addition to evaluating the accuracy and relevance of the model's output, human judgment can also guide the model's development. For example, end-user feedback can identify areas where the model's performance could be improved and inform the development of new training data and algorithms.

ChatGPT is a powerful tool for generating high-quality and context-aware responses. Hence, it is essential to understand that the responses generated by the model are not always accurate and require human judgment to be effectively evaluated and interpreted. By monitoring and evaluating the performance of the model and by using human judgment to assess the accuracy and relevance of its output, we can ensure that ChatGPT continues to generate high-quality and trustworthy responses.

CHAPTER 10

The Future of ChatGPT and its Applications

——————◆‹‹◆››◆——————

Chat GPT has come a long way since its inception and continues to evolve and improve with AI and machine learning advancements. The future of ChatGPT is promising, and its applications are expected to expand into new areas, further enhancing its impact on industries, businesses, and everyday life.

In this chapter, we will explore the future of ChatGPT, including the advancements and improvements expected to be made soon. We will also discuss the potential applications of ChatGPT and how these applications are expected to shape the future of industries and businesses.

Advancements in ChatGPT

ChatGPT is a rapidly evolving technology; many advancements and improvements are expected soon. Some of the key advances that are expected to be quickly made include:

1. Improved accuracy and relevance of responses: The accuracy and applicability of ChatGPT's responses are expected to continue to improve as advancements are made in the field of AI and machine learning. This will result in more accurate and relevant responses from

ChatGPT, enhancing its impact on industries, businesses, and everyday life.

2. Enhanced natural language processing: Natural language processing (NLP) is a crucial component of ChatGPT, and advancements in NLP are expected to lead to improved natural language processing capabilities in ChatGPT. This will improve language understanding and enable ChatGPT to generate more natural and human-like responses.

3. Increased ability to handle complex tasks: ChatGPT is expected to handle increasingly tricky tasks soon as advancements are made in machine learning and AI. This will expand the range of applications for ChatGPT and will allow it to tackle more complex challenges.

4. Increased use of deep learning: The use of deep learning is expected to increase in ChatGPT, resulting in improved accuracy and relevance of responses. Deep learning is a powerful tool for improving machine learning algorithms, and its use in ChatGPT will lead to further advancements in the field.

Potential Applications of ChatGPT

The potential applications of ChatGPT are vast and far-reaching and are expected to continue to expand in the future. Some of the potential applications of ChatGPT include:

1. Customer service: ChatGPT is already being used for customer service in various industries, and its use is

expected to expand. ChatGPT can provide quick and accurate responses to customer queries, improving customer satisfaction and reducing the workload for human customer service representatives.

2. Language translation: ChatGPT is expected to play a major role in language translation as it becomes increasingly capable of accurately translating between languages. This will make communication between individuals who speak different languages easier and positively impact industries and businesses that rely on language translation.

3. Content generation: ChatGPT is already being used for content generation and is expected to expand. ChatGPT can generate high-quality content quickly and efficiently, which can be used for various purposes, including marketing, journalism, and creative writing.

4. Healthcare: ChatGPT has the potential to play a major role in the healthcare industry as it becomes increasingly capable of handling complex medical tasks. For example, ChatGPT can assist in medical diagnoses, provide information on medical treatments, and help improve the overall quality of care in the healthcare industry.

Education: Chat G PT has the potential to revolutionize the field of education as it becomes increasingly capable of providing personalized and individualized learning experiences. ChatGPT can be used to provide students with answers to their questions, assist with homework and other assignments, and even provide personalized recommendations for further learning. In addition, ChatGPT can be used to create engaging and interactive learning experiences, which can help to improve student engagement and retention of information.

Financial services: ChatGPT has the potential to play a major role in the financial services industry as it becomes increasingly capable of handling complex financial tasks. For example, ChatGPT can assist with financial planning, provide investment recommendations, and even handle simple financial transactions. This will improve the speed and efficiency of financial services and provide access to financial services to those who may have previously been excluded due to a lack of financial literacy.

Entertainment: ChatGPT has the potential to play a major role in the entertainment industry as it becomes increasingly capable of generating personalized recommendations and experiences. ChatGPT can recommend movies, TV shows, books, and other forms of entertainment based on individual preferences and provide personalized playlists and other recommendations. This will improve the entertainment experience and provide consumers with a more personalized and engaging experience.

These are just a few examples of the potential applications of ChatGPT, and as advancements are made in the field of AI and machine learning, new and innovative applications will likely emerge. It is clear that the future of ChatGPT is bright, and its impact on industries and businesses, as well as everyday life, is expected to be significant.

Since its inception, ChatGPT has come a long way and continues to evolve and improve with advancements in AI and machine learning. The future of ChatGPT is promising, and its applications are expected to expand into new areas, further enhancing its impact on industries, businesses, and everyday life. As ChatGPT continues to evolve, it will likely become an increasingly integral part of our lives, transforming how we work, communicate, and live.

CHAPTER 11
ChatGPT and its Imminent Takeover

---❖❝◆❞❖---

I t's hard to believe, but the future of ChatGPT is here and it's bigger, better, and more advanced than ever before. ChatGPT has come a long way since its early days of stumbling over simple questions and delivering robotic responses. Today, ChatGPT is a true marvel of AI technology, and its impact on our daily lives will only get more pronounced in the future.

ChatGPT is a never-ending journey of self-discovery and self-improvement, striving to be the best version of itself. So, buckle up and get ready for the ride because the future of ChatGPT is about to take us on one wild adventure.

The Imminent Takeover

With all its advancements and potential applications, it's clear that the future of ChatGPT is incredibly bright. It's only a matter of time before ChatGPT becomes so advanced and integrated into our daily lives that it becomes the norm. And that, my friends, is the imminent takeover of ChatGPT.

So, how will this takeover play out? Well, it's hard to say for sure, but one thing is certain - ChatGPT will become a ubiquitous part of our lives, just like smartphones and the internet. Whether helping us with our daily tasks, answering our questions, or entertaining us, ChatGPT will be there for us, becoming an integral part of our daily routine.

An interesting possibility is that of rival developers creating their own versions of ChatGPT. As we all know, this will create competition amongst the big tech companies, and as a result, will rapidly increase the development of AI.

As tech giants such as Google, Microsoft, and Amazon continue to invest heavily in AI, they are pushing the boundaries of what is possible and driving innovation in the field. This competition fosters an environment of collaboration, with researchers and engineers working tirelessly to improve the performance of their respective models.

With each new release of a language model, the competition becomes more intense, as each company strives to create the most powerful and capable AI tool. As a result, we have seen tremendous advancements in the capabilities of ChatGPT, with each iteration surpassing the previous one in terms of its ability to understand and respond to human language.

This competition also creates a market for AI services, which has led to an increase in investment in the development of ChatGPT. This investment has enabled developers to improve the model's performance, add new features, and integrate it into a wide range of applications.

Furthermore, the competition among big tech companies has led to the emergence of the new use cases for ChatGPT. Companies are finding innovative ways to integrate the model into their products and services, creating new revenue streams and driving growth in the industry. As companies continue to invest in an AI and push the boundaries of what is possible, we can expect to see even more significant advancements in the capabilities of ChatGPT.

The future of ChatGPT is here, and it's looking brighter than ever. With its constant evolution, improved capabilities, and endless potential applications, ChatGPT is poised to take over and become the norm in our daily lives. So, get ready for the ride because the future of ChatGPT is about to take us on one wild adventure.

CHAPTER 12

Harnessing the Power of Chatgpt in the Classroom

————————— ◆ ◄◄ ◆ ►► ◆ —————————

As technology continues to play an increasingly important role in our daily lives, it's no surprise that it is also changing how we learn. ChatGPT is one of the most powerful tools available to teachers and students today, and it can potentially revolutionise how we approach education. In this chapter, we will explore how ChatGPT can be used to enhance the learning experience in the classroom.

1. Personalized Learning

One of the biggest benefits of using ChatGPT in the classroom is that it can create personalized student learning experiences. Using machine learning algorithms, ChatGPT can adapt to the needs and abilities of each student, providing them with the resources and support they need to succeed. This allows for a more efficient and effective learning experience, as students can work at their own pace and receive tailored feedback and support.

2. Enhancing Collaboration

ChatGPT can also be used to enhance collaboration in the classroom. By providing students with a platform to communicate with one another and their teacher, ChatGPT makes it easier for students to collaborate on projects, share resources and ideas, and support each other. This can help foster a sense of community in the classroom and lead to a more engaged and motivated learning experience.

3. Supporting Language Learning

For students who are learning a new language, ChatGPT can be a valuable tool. By providing students to practice their language skills in real-time, ChatGPT can help to improve their speaking and writing abilities. It can also give the students feedback and support, helping them identify areas for improvement and grow as language learners.

4. Supplementing Classroom Instruction

ChatGPT can also supplement classroom instruction, providing students additional resources and support outside the classroom. This can include access to study materials, practice exercises, and other learning resources, all of which can help students to deepen their understanding of the material and to make more meaningful connections with the content.

5. Enhancing Assessment

Finally, ChatGPT can also be used to enhance assessment in the classroom. By providing teachers with real-time insights into student understanding, ChatGPT can be used to inform and improve the assessment process. This can include automated grading and feedback, saving teachers time and providing students with more meaningful feedback.

Implementing ChatGPT in the Classroom

While the benefits of using ChatGPT in the classroom are clear, there are some steps that teachers can take to ensure that they are making the most of this powerful tool. Here are a few tips to help you get started:

1. Start Small

When implementing ChatGPT in the classroom, starting small and working your way up is essential. Try using ChatGPT to support a single aspect of your curriculum, such as language learning or assessment, before expanding to other areas. This will help you get a feel for how ChatGPT works and what it can do and allow you to make any necessary adjustments along the way.

2. Engage Students

It's essential to engage students in using ChatGPT and ensure they understand how it can support their learning. Please encourage students to use ChatGPT for collaborative projects, language practice, and additional learning resources, and provide them with opportunities to give feedback on their experience.

3. Collaborate with Other Teachers

Finally, it's essential to collaborate with other teachers when implementing ChatGPT in the classroom. Share your experiences and best practices with your colleagues, and work together to develop ways to integrate ChatGPT into your curriculum. This can include developing joint projects or resources or hosting professional development workshops to help other teachers get started with ChatGPT.

Evaluate and Refine As you implement ChatGPT in the classroom, it's essential to evaluate its impact on student learning regularly and make necessary adjustments. This can include surveying students and teachers to gather feedback on their experience, tracking student progress and engagement, and refining your approach as needed. By regularly evaluating and refining your use of ChatGPT, you can ensure that you are making the most of this powerful tool and that your students get the best possible learning experience.

In conclusion, ChatGPT can revolutionise how we approach education and provide students with a more personalized, engaging, and effective learning experience. Whether you are a teacher looking to integrate new technology into your curriculum or a student looking to get the most out of your education, ChatGPT is a tool you won't want to miss.

Let's look at the education angle a bit differently. I will include different points of view about ChatGPT in the classroom; which one do you agree with?

Point of View 1: The Enthusiastic Teacher- As an enthusiastic teacher, ChatGPT can be a game-changer in the classroom. Its ability to provide instant feedback and personalized learning experiences can help students achieve better learning outcomes. For instance, ChatGPT can be used to develop interactive educational games that make learning more engaging and enjoyable for students. It can also be used to create virtual tutors that can assist students in their studies by providing them instant feedback and guidance.

Point of View 2: The Cautious Educator- As a cautious educator, I understand the potential of ChatGPT in the classroom but also recognize the need to use it in a controlled and thoughtful manner. One potential concern is the reliability of the information provided by ChatGPT, as it may not always be accurate or relevant to the specific learning goals. Additionally, there is a risk of over-reliance on ChatGPT, which may lead to students lacking critical thinking and problem-solving skills. Therefore, it's essential to carefully consider the use cases of ChatGPT in the classroom and balance it with other forms of learning.

Point of View 3: The Technology Skeptic- As a technology skeptic, I am skeptical about the benefits of ChatGPT in the classroom. While it may provide instant feedback and personalized learning experiences, it can also lead to a lack of human interaction and personal touch in the learning process. Additionally, it may not be accessible to all students, particularly those from underprivileged backgrounds who may not have the required technology. Therefore, it's essential to weigh the benefits and drawbacks of ChatGPT in the classroom and ensure that it doesn't replace traditional forms of learning.

Point of View 4: The Progressive Educator- As a progressive educator, ChatGPT can be a valuable addition to the classroom if used to complement other forms of learning. By providing personalized learning experiences and instant feedback, ChatGPT can help students achieve better learning outcomes and foster critical thinking and problem-solving skills. It can also be used to provide accessibility and support to students who may have special needs or face learning challenges. However, it's important to remember the ethical considerations and potential biases arising from using ChatGPT in the classroom.

ChatGPT in the classroom can potentially revolutionise how we approach education. However, it's essential to weigh the benefits and drawbacks of ChatGPT and use it thoughtfully and carefully. By doing so, we can harness the power of ChatGPT to provide personalized and engaging learning experiences that can help students achieve better learning outcomes.

CHAPTER 13

The Role of ChatGPT in Advancing Environmental Sustainability

————————— ◆ ◄◄ ◆ ►► ◆ —————————

As concerns about climate change and environmental degradation continue to grow, it is more important than ever for individuals and organizations to take action to promote sustainability. ChatGPT, a powerful language model developed by OpenAI, is one such technology that can potentially play a significant role in advancing environmental sustainability. In this chapter, we will explore how ChatGPT can be used to address environmental challenges and promote sustainable practices.

Environmental Education, One of the most effective ways to promote environmental sustainability, is through education, and ChatGPT can play a key role in this effort. ChatGPT can help individuals better understand environmental issues and make informed decisions about their actions by providing access to educational resources and information. This can include providing information on climate change, renewable energy, waste reduction, and sustainable living practices.

Sustainable Supply Chain Management: ChatGPT can contribute to environmental sustainability through its use in supply chain management. With its ability to analyze vast amounts of data, ChatGPT can help businesses and organizations to identify areas where they can reduce waste, improve energy efficiency, and minimize their environmental impact. This can include analyzing data related to transportation and logistics, energy usage, and waste management.

Natural Language Processing for Environmental Monitoring Natural language processing (NLP) is a key aspect of ChatGPT's functionality, and it can be used to advance environmental monitoring and conservation efforts. NLP can analyze large volumes of text-based data, such as satellite imagery, social media posts, and news articles, to detect patterns and trends related to environmental issues. This can help researchers and organizations to identify areas of concern and to take action to address them.

Environmental Decision-Making As environmental challenges become more complex, there is a growing need for tools and technologies to support decision-making. ChatGPT can generate models and predictions about environmental scenarios, allowing individuals and organizations to make more informed decisions about the best action. This can include modeling the potential impacts of policy changes, predicting the outcomes of different environmental interventions, and identifying the most effective solutions to complex environmental challenges.

Enhancing Energy Efficiency ChatGPT can also advance energy efficiency and promote sustainable energy practices. By analyzing data related to energy usage, ChatGPT can help individuals and organizations to identify areas where energy can be conserved, such as by turning off lights and electronics when not in use, adjusting thermostat settings, and optimizing building design. This can help to reduce energy consumption and promote more sustainable energy practices.

Using ChatGPT for Environmental Communication Finally, ChatGPT can improve communication around environmental issues, helping to raise awareness and promote action. By generating compelling and engaging content, ChatGPT can help to engage audiences and inspire them to take action. This can include creating content for social media, blogs, and other online platforms and generating educational materials for schools, businesses, and other organizations.

Implementing ChatGPT for Environmental Sustainability

While the potential benefits of using ChatGPT for environmental sustainability are straightforward, several steps must be taken to implement this technology effectively. Here are a few tips to help organizations and individuals get started:

One of the key areas of focus is energy efficiency, which involves reducing energy consumption and optimizing energy use. Organizations can use ChatGPT to identify energy inefficiencies and recommend solutions that can help reduce energy consumption. For instance, ChatGPT can provide employees with energy-saving tips or suggest upgrades to energy-efficient appliances and lighting systems.

Another key area of focus is waste reduction, which involves minimizing the amount of waste generated and increasing the recycling and composting rate. ChatGPT can help organizations develop waste reduction strategies by providing suggestions on reducing waste generation, recycling best practices, and composting tips. ChatGPT can also assist in tracking and reporting waste reduction progress to ensure the organisation meets its targets and objectives.

Sustainable supply chain management is another important area of focus. ChatGPT can be used to optimize the supply chain, reduce waste, and improve resource efficiency. For instance, ChatGPT can recommend sourcing materials from sustainable suppliers, minimising transportation costs, and reducing packaging waste. ChatGPT can also assist in monitoring and reporting on the sustainability of the supply chain, which can help organizations identify opportunities for improvement.

Environmental education is another key area of focus for implementing ChatGPT for environmental sustainability. ChatGPT can educate employees, customers, and other stakeholders about the importance of sustainability and provide information on reducing their environmental footprint. ChatGPT can also promote sustainable behavior and encourage individuals to take action on environmental issues.

Implementing ChatGPT for environmental sustainability requires identifying key focus areas, developing targeted strategies and initiatives, and using ChatGPT to optimize sustainability performance. With its ability to generate data-driven insights, provide recommendations, and automate processes, ChatGPT can help organizations achieve their sustainability goals and positively impact the environment.

Providing Green Energy Solutions

With its ability to analyze large amounts of data and identify patterns, ChatGPT can transform how businesses approach clean energy solutions. One key area where ChatGPT can be used is in developing new and innovative sources of green energy. ChatGPT can help businesses identify the most promising locations for solar panels, wind turbines, and other renewable energy sources by analysing weather patterns, energy consumption, and other factors. This analysis can also help businesses optimize the performance of these energy sources, increasing their efficiency and effectiveness.

Another way ChatGPT can contribute to green energy solutions is by predicting future trends and anticipating potential problems. For example, ChatGPT can forecast changes in energy demand or shifts in energy markets, helping businesses adapt and stay ahead of the curve. It can also identify potential risks or vulnerabilities in energy infrastructure, allowing businesses to take preventative action before problems arise.

ChatGPT can also play a role in educating consumers about the benefits of green energy. By analyzing consumer behavior and preferences data, ChatGPT can help businesses develop targeted marketing campaigns and educational materials encouraging consumers to adopt more sustainable energy practices. Overall, the use of ChatGPT in providing green energy solutions has the potential to drive significant progress towards a more sustainable and environmentally-friendly future.

Reducing Carbon Footprint

Reducing carbon footprint is one of the most important steps towards achieving environmental sustainability. ChatGPT can be a powerful tool to help businesses in this process. The technology can analyze a business's entire supply chain, from raw materials to finished products, to identify areas where energy usage and waste can be reduced. It can also suggest alternatives to traditional manufacturing processes that are more environmentally friendly. ChatGPT can monitor and analyze a business's energy consumption and emissions in real-time, allowing quick action to reduce its carbon footprint.

ChatGPT can also help businesses develop and implement sustainability goals, track progress, and optimize strategies to achieve those goals. By using ChatGPT to reduce their carbon footprint, businesses can improve their environmental impact and benefit financially by reducing energy costs and improving efficiency.

Improving Waste Management

Waste management is a critical issue for environmental sustainability, and ChatGPT can be used to develop more effective waste management strategies. By analyzing data on waste production, ChatGPT can help businesses identify the sources of waste and the types of waste being produced. It can also be used to identify more efficient and effective ways to dispose of waste, such as through recycling or composting. Businesses can significantly reduce their environmental impact by reducing waste and improving waste management practices.

Increasing Sustainability Awareness

Finally, ChatGPT can increase awareness and education about environmental sustainability. ChatGPT can help raise awareness about human activity's impact on the environment by providing users with real-time information and feedback on their actions and behaviors. It can also provide information and resources on sustainable practices and solutions, empowering individuals and businesses to take action to protect the environment.

Implementing ChatGPT for Environmental Sustainability

Implementing ChatGPT for environmental sustainability can be a complex and challenging process, but there are several steps that businesses can take to ensure success:

Identify Goals and Objectives

The first step in implementing ChatGPT for environmental sustainability is identifying clear goals and objectives. Businesses should assess their current environmental impact and identify areas for improvement. They should develop specific, measurable goals and objectives aligning with their sustainability strategy.

Develop a Data Strategy

ChatGPT relies on data to function effectively, so businesses should develop a clear data strategy that identifies the data sources and types needed. This may include data on energy consumption, waste production, and emissions, as well as data on weather patterns, market trends, and other factors that can impact sustainability.

Engage Stakeholders

Implementing ChatGPT for environmental sustainability requires buy-in and engagement from all stakeholders, including employees, customers, and suppliers. Businesses should communicate the benefits of sustainability and ChatGPT and encourage stakeholders to participate actively in the implementation process.

Monitor and Evaluate Progress

Finally, businesses should monitor and evaluate their progress regularly to ensure they achieve their sustainability goals. They should use ChatGPT to track and analyze data and to identify areas where improvements can be made. By continuously evaluating their progress and making adjustments as needed, businesses can ensure they are on the path to long-term environmental sustainability.

Furthermore, ChatGPT can also help businesses to achieve their sustainability goals by analyzing customer data and identifying opportunities to promote more sustainable products and practices. For instance, machine learning algorithms can analyze consumer behavior and preferences to determine which products are most popular and sustainable. This can help businesses to develop more sustainable products and to promote them effectively to consumers.

Implementing ChatGPT in Business

If you are interested in implementing ChatGPT in your business to improve your environmental impact, remember a few things. Here are some tips to help you get started:

1. Identify Your Objectives: The first step in implementing ChatGPT in your business is to identify your objectives. What are your environmental goals, and how can ChatGPT help you to achieve them? Once you clearly understand your objectives, you can explore the specific ways ChatGPT can be applied to your business.

2. Gather Data: To make the most of ChatGPT, you must have access to high-quality data. This may include data from sensors, monitoring systems, customer data, and other sources. Ensure you understand the data you need, where it comes from, and how to access it.

3. Choose the Right Tools: Many different ChatGPT tools are available, each with strengths and weaknesses. Make sure you choose a tool well-suited to your specific needs and objectives.

4. Involve Your Team: Implementing ChatGPT in your business is a team effort. Make sure that you involve your team in the process and that they understand how ChatGPT works and how it can help to improve your environmental impact.

5. Measure Your Results: Finally, measuring and tracking your progress over time is essential. This will help you identify improvement areas and ensure you achieve your environmental objectives.

Conclusion

ChatGPT is a powerful tool that has the potential to revolutionize the way we approach environmental monitoring and sustainability in business. By analyzing large volumes of data and identifying patterns and opportunities for improvement, ChatGPT can help businesses to reduce their environmental impact, save money, and improve customer satisfaction. If you want to use ChatGPT to improve your environmental impact, there has never been a better time to get started. With the right tools and a clear understanding of your objectives, you can make a real difference and help to create a more sustainable future for us all.

CHAPTER 14

The Future of AI Software Developed from ChatGPT

————————— ◆ ◄◄ ◆ ►► ◆ —————————

As we have seen, ChatGPT has already revolutionized the way we interact with AI software. However, the potential for ChatGPT goes far beyond its current capabilities. In this chapter, we will explore the future of AI software that can be developed from ChatGPT and the impact it could have on various industries.

Natural Language Processing

One area where we can expect to see major developments soon is natural language processing (NLP). NLP studies how computers can be programmed to understand and interpret human language. ChatGPT has already made significant strides in this area, but there is still much room for improvement.

In the future, we can expect to see AI software that is even better at understanding the nuances of human language. This could lead to more accurate and sophisticated chatbots and improved machine translation and sentiment analysis. Businesses could benefit from this technology by using it to improve customer service, marketing, and product development.

Machine Learning

Perhaps the most underrated use of ChatGPT is Machine Learning. Machine learning is a subset of artificial intelligence that has recently gained significant attention and popularity. With the rise of big data and the need to process vast amounts of information, machine learning has become an essential tool for businesses and organizations across all industries. ChatGPT, with its ability to learn from vast amounts of text data, is poised to significantly impact the machine learning process.

These algorithms can be broadly categorized into three types: supervised learning, unsupervised learning, and reinforcement learning. In supervised learning, the algorithm learns from labeled data to make predictions or classifications on new, unlabeled data. On the other hand, unsupervised learning involves finding patterns or relationships in unlabeled data without specific guidance or training. Reinforcement learning is a trial-and-error process in which an algorithm learns by interacting with an environment and receiving feedback or rewards for its actions. With the help of ChatGPT, machine learning algorithms can become even more accurate and efficient. ChatGPT can be used to preprocess text data, clean and transform data, and even generate new data for training purposes.

ChatGPT's ability to understand and generate natural language can also help improve the interpretability and explainability of machine learning models. This is especially important in applications such as healthcare or finance, where decisions based on machine learning algorithms need to be transparent and explainable. Furthermore, ChatGPT can also be used for predictive modeling, such as forecasting demand, predicting customer behavior, or identifying potential risks. Using machine learning models powered by ChatGPT, businesses can make better decisions, optimize operations, and drive growth and profitability. The potential applications of ChatGPT in machine learning are vast, and we can expect to see more and more businesses adopting this technology in the coming years.

Virtual Assistants

Virtual assistants have become increasingly popular recently, with many people using them to manage daily tasks and schedules. However, current virtual assistants are still relatively limited in their capabilities. ChatGPT has the potential to change this.

In the future, we can expect to see virtual assistants that are much more sophisticated and capable. These virtual assistants could be programmed to understand and interpret natural language, making them more responsive to user needs. They could also be used to automate many tasks that are currently done manually, such as scheduling appointments or managing emails. Businesses could benefit from this technology to improve productivity, streamline operations, and enhance customer service.

Robotics

Finally, ChatGPT has the potential to significantly impact the field of robotics by improving the way robots are programmed and controlled. While robots are programmed to perform specific tasks, they cannot adapt to new situations and tasks. Using ChatGPT, robots could be programmed to generate responses to new situations in real time, making them more versatile and capable of performing a more comprehensive range of tasks.

One of the main benefits of using ChatGPT in robotics is the ability to generate human-like commands for robots. This means that robots could be programmed to respond to natural language inputs and understand the context of those inputs. For example, a robot could be trained to respond to a request to "pick up the blue box on the shelf" and understand the specific location of the box and the object's color. This level of understanding and responsiveness would make robots much more versatile and valuable in various industries.

In addition to improving the capabilities of robots, ChatGPT could also help to improve workplace safety. Robots are often used in industries such as manufacturing and construction, where accidents and injuries can occur. By programming robots with ChatGPT, they could be trained to recognize and respond to potential safety hazards, such as obstacles in their path or unsafe working conditions.

Integrating ChatGPT in robotics could lead to significant advancements in the field, resulting in more versatile, capable, and intelligent robots that can perform various tasks. Businesses in manufacturing, construction and logistics industries could benefit from this technology by improving operational efficiency, increasing productivity, and enhancing workplace safety.

Conclusion

ChatGPT has already had a major impact on AI software, and its potential for the future is immense. With further advancements in natural language processing, machine learning, virtual assistants, and robotics, we can expect to see AI software that is even more sophisticated and capable than what we have today. The possibilities are endless, and businesses that can harness the power of ChatGPT and its potential for the future will be well-positioned to succeed in the years to come.

CHAPTER 15

ChatGPT and Its Potential to Help Senior Citizens

―――――――――◆ ◄◄ ◆ ►► ◆――――――――

Senior citizens often face challenges such as loneliness, social isolation, cognitive decline, and physical limitations. ChatGPT has the potential to address these challenges and improve the quality of life for seniors in many ways.

ChatGPT can help seniors by providing companionship and social interaction. Many seniors live alone, and social isolation is a significant problem that can lead to depression and other mental health issues. ChatGPT can be a virtual companion to converse with seniors and provide a sense of connection and companionship. It can also keep them engaged and mentally stimulated, improving cognitive function and preventing mental decline.

As the population continues to age, the demand for healthcare services increases. For seniors, managing chronic health conditions can be a daunting task. ChatGPT can play a vital role in assisting seniors with their healthcare needs by providing personalized health information and advice.

ChatGPT can create a personalized health profile for seniors, including their medical history, medication schedule, and other health-related information. This profile can be used to provide tailored advice and reminders based on the senior's specific health needs. ChatGPT can also monitor medication schedules and alert seniors when it's time to take their medications.

In addition to medication management, ChatGPT can provide seniors with information about their health conditions and offer advice on managing their symptoms. This can include dietary recommendations, exercise suggestions, and other lifestyle changes to help seniors manage their needs more effectively.

Moreover, ChatGPT can also be used to connect seniors with medical professionals if necessary. This can be particularly useful for seniors who live in rural or remote areas where access to healthcare services may be limited. ChatGPT can guide when to seek medical attention and connect seniors with healthcare professionals via telemedicine services.

ChatGPT can also be beneficial to seniors with hearing or vision impairments. It can transcribe speech in real-time and convert it to text, which can help seniors communicate more effectively. It can also read text messages or emails to seniors struggling to read independently.

Also, ChatGPT be used to create virtual reality experiences that can simulate travel, education, and entertainment. This can be especially useful for seniors with physical limitations that prevent them from traveling or participating in activities outside the home.

While ChatGPT has enormous potential to help seniors, challenges must be addressed. One concern is the potential for seniors to become too reliant on technology, leading to a lack of social interaction and physical activity. Another issue is the need to ensure the technology is user-friendly and accessible to seniors who may have difficulty with technology.

The potential to be an excellent resource for seniors by providing companionship, personalized health information, virtual assistance, and virtual reality experiences is immeasurable regarding ChatGPT. It can help seniors overcome social isolation, manage their health conditions, and stay engaged and mentally stimulated. However, it is essential to address potential challenges and ensure that the technology is accessible and user-friendly for seniors. We are barely scratching the surface of how this can change the lives of the elderly.

CHAPTER 16

The Influence of ChatGPT on Other Technologies

———————◆◄◀◆►►◆———————

S ince its inception, ChatGPT has become a trailblazer in artificial intelligence. Its sophisticated natural language processing capabilities have enormously impacted other technological innovations, transforming how we interact with machines. In this chapter, we will explore how ChatGPT has influenced various other tech fields and how it continues to shape the future of AI.

Voice recognition is one of the most significant areas where ChatGPT has impacted other technologies. Voice assistants like Alexa and Siri have become ubiquitous in our daily lives, and ChatGPT has played an essential role in their development. By processing vast amounts of natural language data, ChatGPT has helped these virtual assistants become more responsive and accurate in their responses. They can now understand complex requests, thanks in part to the language processing capabilities of ChatGPT.

While ChatGPT has helped improve the capabilities of various technologies, it has also raised important ethical concerns about the use of AI in society. As AI-powered technologies become increasingly prevalent daily, the potential for bias and discrimination becomes a growing concern. ChatGPT's ability to generate realistic language prompts raises questions about the ethics of using AI to create fake news or manipulate public opinion.

Despite these concerns, ChatGPT also has the potential to be used for good. For instance, it can help improve accessibility for individuals with disabilities. ChatGPT can assist with speech-to-text transcription, making it easier for those with hearing impairments to follow along with conversations. It can also help those with speech impairments communicate more easily with others.

ChatGPT is a powerful tool that has significantly contributed to various technological fields. Its natural language processing capabilities have improved communication and enabled researchers to develop more advanced AI models. However, there are also ethical concerns regarding the potential misuse of this technology, particularly in generating fake content or manipulating public opinion.

Despite these concerns, ChatGPT has the potential to greatly benefit society by making communication more accessible and user-friendly. It can help people with disabilities or language barriers to connect more easily with others and access information. Furthermore, ChatGPT can assist businesses in optimizing their operations, improving customer engagement, and making more informed decisions.

As the technology continues to evolve, we can expect to see even more applications of ChatGPT in various fields, including healthcare, education, and environmental sustainability. Integrating ChatGPT with other technologies will lead to more advanced and innovative solutions. However, it is essential to carefully consider the ethical implications of using ChatGPT and ensure that it is being used responsibly and beneficially.

CHAPTER 17

The Power of ChatGPT in Modern Warfare

————————◆ ◀◀◆▶▶◆————————

Let's get a bit more serious, shall we?

As the development of AI continues to advance rapidly, it's no surprise that it has also had a significant impact on the world of modern warfare. With ChatGPT, military leaders now have access to a powerful tool that can process and analyze large amounts of data in real-time, providing valuable insights to help them make more informed decisions on the battlefield.

ChatGPT's ability to process and analyze vast amounts of data has revolutionized modern warfare. Military operations rely heavily on data, and ChatGPT's natural language processing capabilities allow it to understand and analyze historical military data to identify patterns and potential weaknesses in enemy tactics.

By training on this data, ChatGPT can provide valuable insights into the strategies and tactics used by enemy forces, allowing military leaders to develop effective countermeasures. For example, ChatGPT can analyze data on enemy troop movements and identify potential vulnerabilities in their formations, allowing military planners to develop targeted attacks.

In addition to its ability to analyze historical data, ChatGPT can process real-time data from various sources, including satellite imagery, drone footage, and social media. This real-time data can provide military leaders valuable insights into enemy movements and activities, allowing them to adjust their strategies and tactics accordingly.

Moreover, ChatGPT can assist in developing autonomous weapons systems capable of analyzing data in real-time and making decisions based on that data. This could reduce the need for human intervention in combat situations, leading to a more efficient and effective military force.

Moreover, ChatGPT's advanced natural language processing capabilities have led to the development of sophisticated chatbots and virtual assistants (as discussed previously) that can provide real-time updates and recommendations to soldiers on the ground. These AI-powered assistants can help soldiers make more informed decisions and stay ahead of the enemy, potentially saving lives.

However, as with any technology, there are ethical concerns surrounding ChatGPT in warfare. For instance, using AI-powered weapons raises questions about the role of human oversight and the potential for autonomous decision-making. The use of ChatGPT-generated language prompts for psychological warfare and propaganda purposes is also a growing concern.

Despite these concerns, the integration of ChatGPT into modern warfare is already well underway, and the potential for this technology is vast. With its ability to analyze large volumes of data in real-time, military leaders can make more informed decisions, potentially saving lives and minimizing the impact of conflict on civilians.

ChatGPT has significantly impacted modern warfare, from its ability to process and analyze large amounts of data to developing advanced chatbots and virtual assistants. While there are ethical concerns around the use of AI in the military, the potential for this technology is vast, and it will continue to shape the future of warfare and military strategy.

CHAPTER 18

The Future of AI and Its Impact on Jobs Worldwide

---◆◄◄◆►►◆---

Artificial Intelligence (AI) and ChatGPT have changed the world as we know it, and the future of these technologies is set to impact the job market on a global scale. In this chapter, we'll explore the potential impact of AI and ChatGPT on jobs, industries, and regions worldwide. We'll also examine the ethical implications of these new technologies and what they mean for the future of work.

Automation and Job Losses

The impact of AI and ChatGPT on the job market has been a topic of concern and debate. Automation, driven by AI-powered machines, is one of the most significant ways technology has affected employment. As machines become more advanced, they can perform tasks previously done by humans, leading to job losses in some industries. For example, in manufacturing, machines can perform repetitive tasks with greater efficiency and accuracy, leading to the loss of jobs in the sector.

However, the impact of automation on the job market is not entirely negative. Some jobs may become obsolete, but new roles will emerge that focus on managing and maintaining these machines and developing new AI technologies. This means that workers must adapt their skills to meet the demands of the changing job market. It is also worth noting that the automation of specific tasks can increase productivity and efficiency, leading to cost savings and economic growth.

It is important to note that not all jobs are equally susceptible to automation. Jobs that require human intuition, creativity, and empathy are less likely to be automated. For example, careers in healthcare, education, and the arts are less likely to be impacted by automation. However, even in these industries, AI and ChatGPT can be used to enhance and improve the quality of work. For example, AI can help healthcare professionals to make more accurate diagnoses and personalize treatments, ultimately improving patient outcomes.

Emerging Industries and Job Roles

AI and ChatGPT are driving the emergence of entirely new industries and job roles. With the increasing integration of AI-powered technologies into our daily lives, the demand for skilled workers in data analysis, machine learning, and natural language processing is rising. As these technologies evolve, new opportunities for those with the skills to develop and maintain them will arise.

In addition to creating new jobs in the technology industry, AI technologies are also revolutionizing our work. The global pandemic has accelerated the adoption of remote work, and AI-powered tools enable people to work from anywhere. This has opened up new opportunities for people in rural areas, where traditional job opportunities may be limited. Remote work also provides greater flexibility and work-life balance, which can be especially beneficial for working parents and individuals with disabilities.

Furthermore, as AI technologies become more integrated into various industries, they create new job roles previously unavailable. For example, the healthcare industry is now seeing a rise in demand for medical data analysts who can use AI to extract insights from patient data. Similarly, the retail sector uses AI to optimize inventory management and enhance customer experience, creating new roles for data analysts and machine learning experts.

However, the emergence of these new industries and job roles also means that workers must adapt and acquire new skills to remain competitive in the job market. AI-powered technologies may replace traditional jobs, but new jobs require different skill sets. This means that workers must be proactive in learning new skills and continuously improving their knowledge and expertise.

Industry-Specific Impacts

The impact of AI and ChatGPT on the job market won't be felt equally across all industries and regions. While some sectors may see increased demand for skilled workers, others may experience job losses due to automation. For example, the financial services industry has already seen significant changes due to AI and ChatGPT, with the rise of robo-advisors and algorithmic trading.

In contrast, industries such as healthcare and education are expected to see increased demand for skilled workers as AI technologies are integrated into these sectors. For example, AI-powered medical diagnosis tools can help doctors make more accurate diagnoses, while AI-powered chatbots can provide personalized tutoring to students.

Ethical Implications

As machines become more advanced, they may be able to perform tasks previously considered skilled labor. This raises questions about the value of human labor and the potential for job displacement on a large scale. It's essential to consider the ethical implications of these new technologies and work towards creating a more equitable and accessible future for all.

Additionally, there are concerns about bias in AI-powered hiring processes. AI technologies can help eliminate discrimination in the hiring process by removing human tendencies, but they can also perpetuate existing biases if not developed and tested correctly. Ensuring these technologies are developed and used responsibly is essential to provide a fair and unbiased job market.

Artificial intelligence will undoubtedly change the job market in ways we can't imagine. While automation may lead to job losses in some industries, it will also create new opportunities for skilled workers in others. As workers adapt to the demands of the changing job market, it's essential to consider the ethical implications of these new technologies and work towards creating a more equitable and accessible future for all. The future is uncertain, but one thing is clear: AI and ChatGPT will shape the job market for years.

CHAPTER 19

The Use of ChatGPT in Mental Health Therapy

———————— ◆ ◂◂ ◆ ▸▸ ◆ ————————

Mental health therapy can be a daunting and costly experience. With long waiting lists and limited availability, it's no wonder many people don't seek the help they need. But what if there was a way to access therapy that was convenient, accessible, and anonymous? Enter ChatGPT, the AI-powered language model revolutionising how we approach mental health therapy.

ChatGPT is used in various applications, from customer service to writing assistance, but its application in mental health therapy is particularly noteworthy. Virtual therapists powered by ChatGPT can now provide text-based therapy sessions to individuals seeking support and guidance in managing their mental health and well-being.

The anonymity provided by ChatGPT's text-based therapy sessions is one of the most significant benefits. For individuals who may be hesitant to seek help due to the stigma surrounding mental health issues, text-based therapy can provide a safe and judgment-free space to express themselves and work through their emotions. With no need to travel or sit in a waiting room, ChatGPT's virtual therapy sessions offer unprecedented accessibility and convenience.

Moreover, ChatGPT's natural language processing capabilities enable it to provide personalized support tailored to individual needs and preferences. Virtual therapists powered by ChatGPT can analyze text responses and provide appropriate and personalized feedback that may be more effective than a scripted response from a human therapist.

It's essential to note that ChatGPT is not a replacement for in-person therapy. Instead, it can be an excellent addition to traditional treatment or an alternative for individuals who cannot access traditional therapy due to cost or lack of availability. However, some experts believe that text-based therapy may be particularly useful for individuals with specific conditions, such as social anxiety, as it allows them to express themselves freely and overcome barriers to seeking help.

While ChatGPT in mental health therapy is still in its early stages, preliminary research has shown promising results. A study published in the Journal of Medical Internet Research found that individuals who received text-based therapy reported significant improvements in their mental health symptoms, including reduced anxiety and depression.

However, concerns surrounding the use of ChatGPT in mental health therapy do exist. One issue is the risk of relying too heavily on technology and not receiving the human connection that is often crucial in therapy. Ensuring that individuals receive adequate support and guidance and that virtual therapists powered by ChatGPT are not used as a substitute for human interaction is essential.

Another potential issue is the limitation of text-based communication. Some individuals may find it challenging to express themselves fully through text-based conversations, which could limit the effectiveness of the therapy. However, this can be overcome by using voice or video calls, which allow for more expressive communication.

The use of ChatGPT in mental health therapy shows significant promise in making therapy more accessible, affordable, and convenient for individuals around the world. With its natural language processing capabilities and personalized feedback, ChatGPT is poised to become an essential tool in the mental health toolbox. While the use of technology in therapy is not without its challenges, it's exciting to see how ChatGPT and other AI-powered tools will continue to transform the mental health landscape.

CHAPTER 20

Maximize Your ChatGPT Experience

◆ ◀◀ ◆ ▶▶ ◆

N ow that you know the benefits of using ChatGPT, it's time to learn how to maximize your experience and get the most out of this fantastic tool. Here are some tips to help you make the most of your ChatGPT conversations:

1. Be specific with your requests: The more specific you are about your request or question, the more accurate and helpful the ChatGPT response will be. Instead of asking a general question, try to be as specific as possible. For example, instead of asking, "How can I be happier?" Try asking, "What strategies can I use to boost my mood daily?"

2. Provide context: ChatGPT works best when it clearly understands your situation and needs. Providing context about your current position or emotions can help ChatGPT tailor its responses to your needs.

3. Use open-ended questions: Instead of asking questions that require a yes or no answer, try asking open-ended questions that encourage a more in-depth response. This will help you get more insight and information from ChatGPT.

4. Experiment with different prompts: ChatGPT responds to various prompts and can offer different types of support, from motivational messages to practical advice.

Experiment with other prompts and see which ones work best for you.

5. Practice active listening: Like in a traditional therapy session, active listening is essential when communicating with ChatGPT. Take the time to read and reflect on ChatGPT's responses, and try to engage with the conversation thoughtfully and meaningfully.

It's also important to remember that ChatGPT is not a replacement for professional therapy or medical advice. If you're struggling with a mental health issue, it's essential to seek the support of a licensed therapist or healthcare professional.

In addition to these tips, you should know a few other things about ChatGPT to make the most of your experience. First, ChatGPT is constantly learning and improving based on user interactions. This means that the more you use ChatGPT, the more accurate and personalized its responses will become.

Second, ChatGPT is powered by advanced natural language processing technology, which means it can understand and interpret human language similarly to how humans communicate. This also means that ChatGPT can pick up on nuances in language and emotions, making it an effective tool for emotional support and guidance.

Finally, it's important to remember that ChatGPT is a tool, not a replacement for human connection and interaction. While it can be a valuable source of support and guidance, it's still essential to seek the help of friends, family, and healthcare professionals when dealing with mental health issues.

By following these tips and understanding the capabilities and limitations of ChatGPT, you can maximize your experience and get the most out of this innovative tool. With ChatGPT, you can access personalized and convenient support to help you improve your mental health and well-being.

CHAPTER 21

Using ChatGPT for Personal Growth and Development

------------◆ ◄◄ ◆ ►► ◆------------

Chat GPT is not just a tool for emotional support; it can also be a valuable resource for personal growth and development. With its ability to understand and interpret human language, ChatGPT can provide users with prompts and exercises to help them develop self-awareness, set goals, and cultivate a growth mindset.

One technique ChatGPT can offer for personal growth and development is the identification of personal strengths. By asking open-ended questions and prompting users to reflect on their past experiences, ChatGPT can help users identify their unique strengths and talents. For example, ChatGPT might ask users to describe when they felt confident and capable and then prompt them to identify the skills and traits that contributed to that experience. Users can build self-confidence and resilience by focusing on their strengths, which are essential for personal growth and development.

Another technique that ChatGPT can offer is the development of self-awareness. ChatGPT can help users become more aware of their patterns and tendencies by asking questions about their thoughts, feelings, and behaviours. For example, ChatGPT might ask users to describe their thought process during a recent challenging experience and then prompt them to reflect on whether their thoughts were helpful or harmful. By developing self-awareness, users can learn to recognize their triggers and develop more effective coping strategies.

ChatGPT can also be a valuable tool for setting and achieving goals. By prompting users to identify their goals and then breaking them down into smaller, actionable steps, ChatGPT can help users stay focused and motivated. For example, ChatGPT might ask a user to identify a long-term goal, such as running a marathon and then prompt them to identify smaller, achievable goals that will help them work towards that goal, such as running a certain distance each week. Users can develop a sense of accomplishment and build momentum towards their larger aspirations by setting and achieving goals.

In addition to these techniques, ChatGPT can be used with other personal growth and development tools. For example, users might combine ChatGPT conversations with journaling or mindfulness practices to deepen their self-awareness and build resilience. Additionally, users might use ChatGPT to set and track goals within a goal-setting app, creating a comprehensive system for personal growth and development.

It's important to remember that while ChatGPT can be a valuable tool for personal growth and development, it's not a replacement for professional therapy or medical advice. If you're dealing with a mental health issue or seeking significant personal growth, it's essential to seek the support of a licensed therapist or healthcare professional.

Here are five ways ChatGPT can help you with your personal growth and development:

1) Personalized goal setting: ChatGPT can assist you in setting personalized goals that align with your interests, strengths, and weaknesses. By engaging in a conversation with ChatGPT, you can discuss your aspirations and receive suggestions on how to achieve them. You can also use ChatGPT to track your progress and receive motivation and encouragement to stay on track.

2) Skill Development: ChatGPT can help you development new skills or improve existing ones. By conversing with ChatGPT about a specific skill, you can receive tips and recourses to enhance your abilities. Additionally, ChatGPT can help you identify areas for improvement and provide feedback to help you grow and develop.

3) Decision-making: ChatGPT has the ability to assist you in making informed decisions by providing insights and information on a topic. Whether you're deciding on a career path or making a personal decision, ChatGPT can help you explore options, weigh pros and cons, and arrive at a well-informed decision.

4) Mindfulness and well-being: ChatGPT can also serve as a mindfulness tool, helping you to reflect on your thoughts and feelings. You can explore emotions, thoughts and beliefs, and gain enlightenment into how they influence your actions and decisions. Also, ChatGPT can provide personalized suggestions to help your well-being, such as tips for managing stress or improving sleep.

5) Personal growth and reflection: It can also help you reflect on your personal growth journey by giving prompts and exercises to explore different experiences and beliefs. By having a conversation with ChatGPT, you can gain clarity on your strengths, weaknesses and areas for growth. ChatGPT can also serve as a source of inspiration, providing quotes and stories to motivate and encourage you on your personal growth journey.

ChatGPT users can develop self-awareness, set and achieve goals, and cultivate a growth mindset. ChatGPT can be a powerful resource for those seeking to reach their full potential when combined with other personal growth and development tools. By incorporating ChatGPT into your personal growth and development plan, you can create a comprehensive system for self-improvement and personal transformation

CONCLUSION

Embracing AI and ChatGPT

————————◆ ◄◄ ◆ ►► ◆————————

As we end this book, we can see that AI and ChatGPT are rapidly becoming integral to our daily lives. From our smartphones to our homes, it's all around us. While this level of integration might seem overwhelming or even scary, it's essential to recognize its potential and be mindful of its limitations and risks.

AI is imperfect, and we cannot rely on it for everything. Machines cannot replace human connection and interaction, which is essential to remember. However, AI can help us in various ways, and we should embrace its potential to revolutionize how we live, work, and learn.

One of the most significant advantages of AI is its ability to process vast amounts of data quickly and accurately. This ability allows us to make informed decisions and solve complex problems that might be beyond the capacity of humans alone. For instance, AI can help doctors diagnose diseases more accurately, assist teachers in personalizing their instruction, and support businesses in identifying trends and predicting future outcomes.

ChatGPT, in particular, has the potential to change the way we communicate with each other. It can help people overcome language barriers, provide mental health support, and even aid personal growth and development. ChatGPT can provide personalized support and guidance, but it's important to remember that it's just a tool. It cannot replace human interaction and empathy, and it's crucial to seek the support of friends, family, and healthcare professionals when dealing with mental health issues.

While AI has the potential to revolutionize various aspects of our lives, we must also be mindful of its limitations and risks. As discussed earlier in this book, AI can be biased and perpetuate systemic discrimination. We must work to ensure that AI is used in a way that is ethical, transparent, and beneficial for society as a whole.

To put a bow on everything, ChatGPT and other forms of AI are not going anywhere, and they have the potential to bring about significant positive changes in our lives. Rather than fear or resist AI, we should embrace it and work to harness its full potential. Doing so can create a brighter and more prosperous future for ourselves and future generations. AI can help us achieve our goals and make our lives easier, but we must also remember that we are responsible for using it to benefit society as a whole. While this book is meant to introduce beginners to the potential of ChatGPT, I hope every person, regardless of their ChatGPT experience, learned something. ChatGPT's future of possibilities is endless, and together, we can master it to create the world of our dreams.

Give other ChatGPT enthusiasts the peace of knowing exactly how incredible and useful AI is.

You have now honed crucial knowledge such as how to use prompt engineering effectively, the many ways to use ChatGPT, the future of Artificial Intelligence, and so much more.

Simply by leaving your honest opinion of this book on Amazon, you'll show other ChatGPT addicts a plethora of practical and important skills that will empower them to feel more confident when using artificial intelligence.

Thank you for your help. You can encourage other people to discover the incredible functions of the revolutionary ChatGPT.

Printed in Great Britain
by Amazon

46775156R00066